THE LANGUAGE OF
STARS & PLANETS

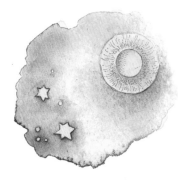

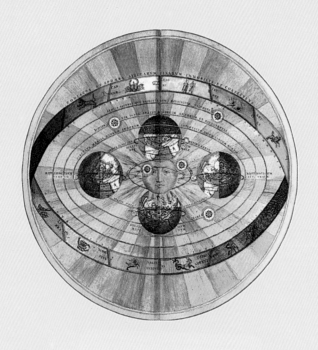

THE LANGUAGE OF
STARS & PLANETS

A VISUAL KEY TO CELESTIAL MYSTERIES

GEOFFREY CORNELIUS & PAUL DEVEREUX

DUNCAN BAIRD PUBLISHERS

LONDON

The Language of Stars & Planets
Geoffrey Cornelius & Paul Devereux

This edition first published in the United Kingdom and Ireland in
2003 by Duncan Baird Publishers Ltd
Sixth Floor, Castle House
75–76 Wells Street
London W1T 3QH

Conceived, created and designed by Duncan Baird Publishers Ltd.

First published in Great Britain in 1996 by Pavilion Books Limited
26 Upper Ground, London SE1 9PD

The authors and publishers thank Sebastian Verney for
permission to use an adaptation of his drawing on page 43.

Editor: Peter Bently
Designer: Gail Jones
Managing Editor: Christopher Westhorp
Managing Designer: Manisha Patel
Commissioned Artwork: David Atkinson, Darren Dubicki,
  Lorraine Harrison
Picture Research: Jan Croot
Science Consultant: Giles Sparrow

British Library Cataloguing-in-Publication Data:
A catalogue record for this book is available from the British
Library

ISBN: 1-904292-63-1

10 9 8 7 6 5 4 3 2 1

Main text typeset in Garamond
Colour reproduction by Colourscan, Singapore
Printed and bound in Thailand by Imago

NOTES
The abbreviations BCE and CE are used throughout this book:
BCE Before the Common Era (the equivalent of BC)
CE Common Era (the equivalent of AD)
Any uncaptioned images are described on page 368.

CONTENTS

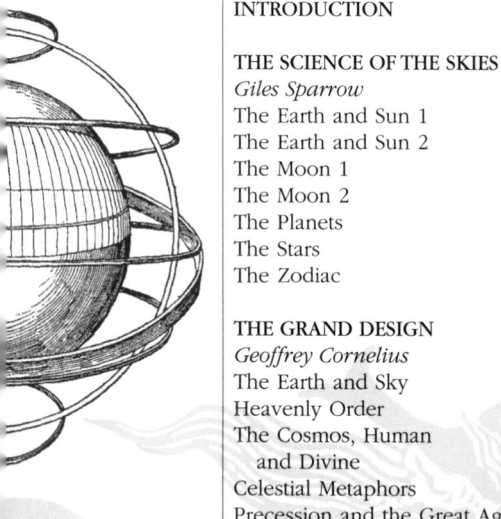

These symbols represent the zodiacal constellations (top) and the Sun, Moon and planets (below). They are derived, in part, from glyphs found on ancient stone reliefs and Egyptian papyri.

♈	♉	♊	♋
Aries	Taurus	Gemini	Cancer
♌	♍	♎	♏
Leo	Virgo	Libra	Scorpio
♐	♑	♒	♓
Sagittarius	Capricorn	Aquarius	Pisces

☉	☿	♀	⊕
Sun	Mercury	Venus	Earth
☽	♂	♃	♄
Moon	Mars	Jupiter	Saturn
♅	♆	♇	
Uranus	Neptune	Pluto	

INTRODUCTION

This book begins with an illustrated account of those aspects of astronomy, and of the technicalities of the zodiac, that are most useful in helping us to understand celestial lore. The language of the stars and planets is also the language of humankind's attempts, in different cultures across the world, to define its place within an ordered cosmos: this is the subject of the second chapter, "The Grand Design". A parallel theme, the evolving relationship between the skies and the realm in which we live, is pursued in the third chapter, "Correspondences", which looks at divination, astrology and the interpretation of celestial symbols and myths in terms of archetypes. The two "Visual Directories", on the "Sun, Moon and Planets" and the "Constellations",

Introduction

narrow the focus by looking at the symbolic associations of some of the most
significant celestial bodies and star-
groups. Finally, in "Sacred Alignments",
our attention shifts to monuments
around the world that show evidence of
a profound impulse in the remote past
to trace connections between the earthly and heavenly realms – to make sightlines to the Sun, Moon and stars at
special times of the year.

A NOTE ON STAR NAMES
The star maps on pages 154–231 of this
book follow astronomical convention in
naming the stars of any given constellation
with the letters of the Greek alphabet.
In most cases, alpha is the brightest star,
beta the next brightest, and so on. Below
are the letters of the Greek alphabet and
their names.

α alpha	ι iota	ρ rho
β beta	κ kappa	σ sigma
γ gamma	λ lambda	τ tau
δ delta	μ mu	υ upsilon
ε epsilon	ν nu	φ phi
ζ zeta	ξ xi	χ chi
η eta	ο omicron	ψ psi
θ theta	π pi	ω omega

A diagram of *c.*1543 showing the orbits of the "ancient" planets around the Sun (the "modern" planets Uranus, Neptune and Pluto were not known at this time) and the band of the zodiac beyond. This view of the Solar System was unthinkable before 1510, when the Polish astronomer Copernicus first proposed that the Earth was not at the centre.

THE SCIENCE
OF THE SKIES

The symbolic associations of the skies and the heavenly bodies cannot be understood without reference to astronomy. Eclipses, the phases of the Moon, the zodiac, key times, such as the equinoxes and solstices, and the notion of great ages, such as the Age of Aquarius, all arise from the Earth, Sun, Moon, planets and constellations shifting their relative positions like a vast and complex machine which, night by night, month by month and year by year, constantly renews the skies we see.

A 19th-century orrery, a mechanical model of the Solar System used to replicate the comparative orbits of the planets around the Sun.

THE EARTH AND SUN 1

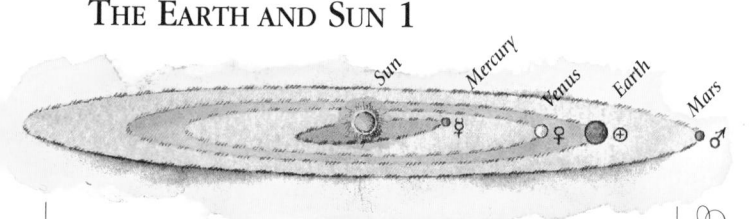

The Sun, the brilliant, fiery star at the heart of our Solar System, controls night and day, the length of the seasons and our measurement of time. It is little wonder that the ancients erected great monuments to celebrate its motions.

In fact, the Sun does not move at all – it only seems to do so because the Earth's movements shift our perspective. The Earth orbits the Sun at an average distance of 93 million miles (150 million km) from it, completing a full orbit in 365.25 days, our solar year: the quarter days add up to make a leap day every fourth year. All the planets orbit the Sun on or close to the same plane, which is imagined as a flat disk centred on the Sun and called the "ecliptic". From the

The four rocky, "terrestrial" planets of the inner Solar System with their conventional symbols: Mercury, closest to the Sun, orbits in 88 days; Venus, which is almost as large as the Earth, orbits in 225 days; the Earth orbits in 365.25 days (one year); and Mars orbits in 687 days.

The Earth and Sun 1

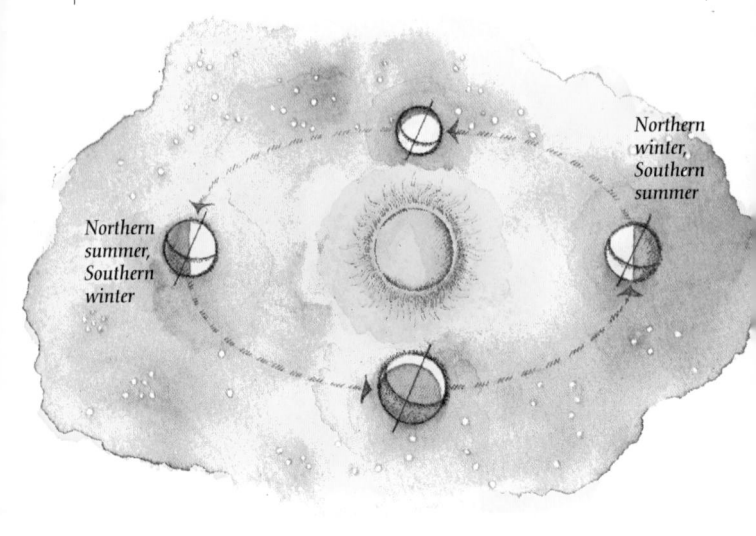

Northern winter, Southern summer

Northern summer, Southern winter

As the Earth orbits the Sun, its axis always points in the same direction. This accounts for the seasons: for half the year the Northern Hemisphere is tipped toward the Sun; for the other half of the year the Southern Hemisphere receives more sunlight.

viewpoint of the Earth, the ecliptic is traced by the path of the Sun as it apparently travels around the sky. As it orbits the Sun, the Earth also rotates on its own axis, an imaginary line through the Earth joining the North and South Poles. A rotation takes 24 hours, during which time any given point on the Earth's surface will turn to face the Sun (day) and then turn to face away from it (night). From Earth, the Sun appears to circle the planet, rising in the east and setting in the west.

The seasons arise because the orientation of the Earth toward the Sun changes during the Earth's orbit. The Earth's axis is tilted at 23.5° from the "upright" (perpendicular to the ecliptic) and always leans in the same direction (see diagram opposite). On June 21 every year, the North Pole leans closest to the Sun – this is the summer solstice in the Northern Hemisphere, when the Sun reaches its highest point in the sky on the longest day of the northern year (midsummer). Simultaneously, the South Pole is at its furthest from the Sun, and this is the winter solstice in the Southern Hemisphere, the shortest day of the southern year (midwinter). Six months later, on December 21, the Earth has moved to the other side of the Sun and the situation is reversed. Midway between the solstices are the equinoxes, when day and night in both hemispheres are of equal length. The vernal (spring) equinox in one hemisphere is the autumnal equinox in the other.

The Earth and Sun 1

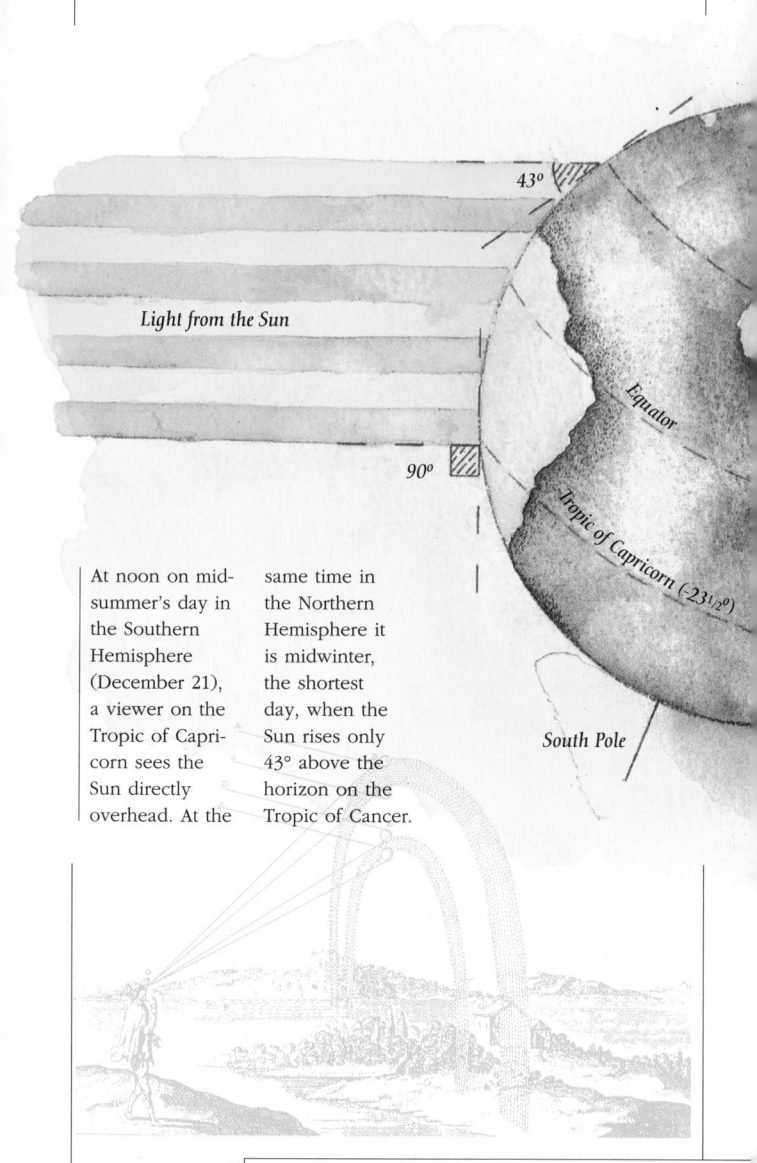

Light from the Sun

43°

90°

Equator

Tropic of Capricorn (23½°)

South Pole

At noon on mid-summer's day in the Southern Hemisphere (December 21), a viewer on the Tropic of Capricorn sees the Sun directly overhead. At the same time in the Northern Hemisphere it is midwinter, the shortest day, when the Sun rises only 43° above the horizon on the Tropic of Cancer.

North Pole

Tropic of Cancer (+23½°)

23½°

THE EARTH AND SUN 2

For an observer on the Earth, our planet's yearly orbit around the Sun creates cycles of apparent movement on the part of celestial bodies. We no longer believe that our planet stands still at the centre of the universe, but for navigating their way around the sky astronomers still use the ancient concept of a "Celestial Sphere", onto which all the objects in the sky are projected, with the Earth at its centre (see diagram, page 37). This sphere is an extension of the Earth's surface into the sky. Any point on the Celestial Sphere follows a circular path in the sky, around the North or South Celestial Pole. The celestial equator – an extension of the

The Earth and Sun 2

Earth's equator into space – divides the Celestial Sphere into two hemispheres.

At the Earth's North Pole, the North Celestial Pole lies directly overhead (that is, at the "zenith", the point on the Celestial Sphere directly above an observer), while at the South Pole the South Celestial Pole is at this point. At the terrestrial equator, the celestial equator passes overhead. The tilt of the Earth's axis to the ecliptic, the plane of the Solar System, means that the celestial equator is at 23.5° to the ecliptic.

The point at which the Sun crosses the celestial equator into the Northern Hemisphere marks the beginning of the Northern spring, called the vernal (spring) equinox, or the "First Point of Aries". From anywhere on Earth, the highest point on the Sun's daily path is when it crosses the "meridian", a line running due south from the zenith (see illustration, pages 22–23). For half of the year the Sun lies in the celestial Southern Hemisphere and for the other half in the Northern. When the Sun moves south of the celestial equator, the days grow shorter in the Northern Hemisphere and the Sun does not rise at the North Pole for six months. Similarly, when the Sun moves north of the celestial equator, days in the Southern Hemisphere grow shorter.

The Earth and Sun 2

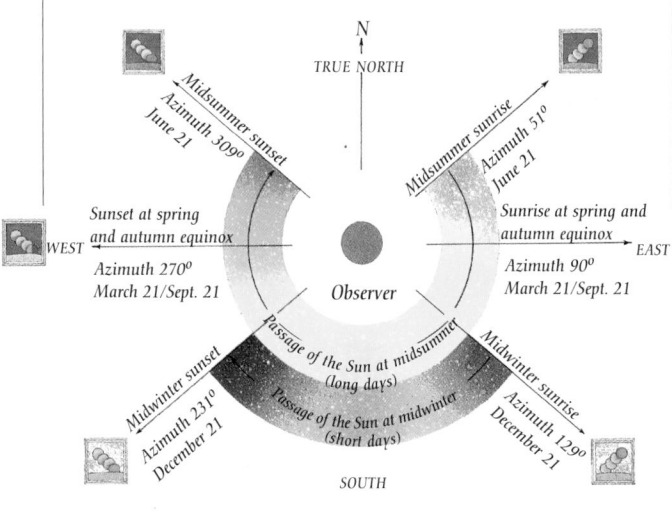

This diagram shows the azimuths (positions on the horizon) of the rising and setting Sun at the solstices and equinoxes at the latitude of Stonehenge (51°N). Azimuth is measured from 0° to 360° clockwise from true north.

ZENITH

Midsummer

Equinox

51°

90°

129°

E

N

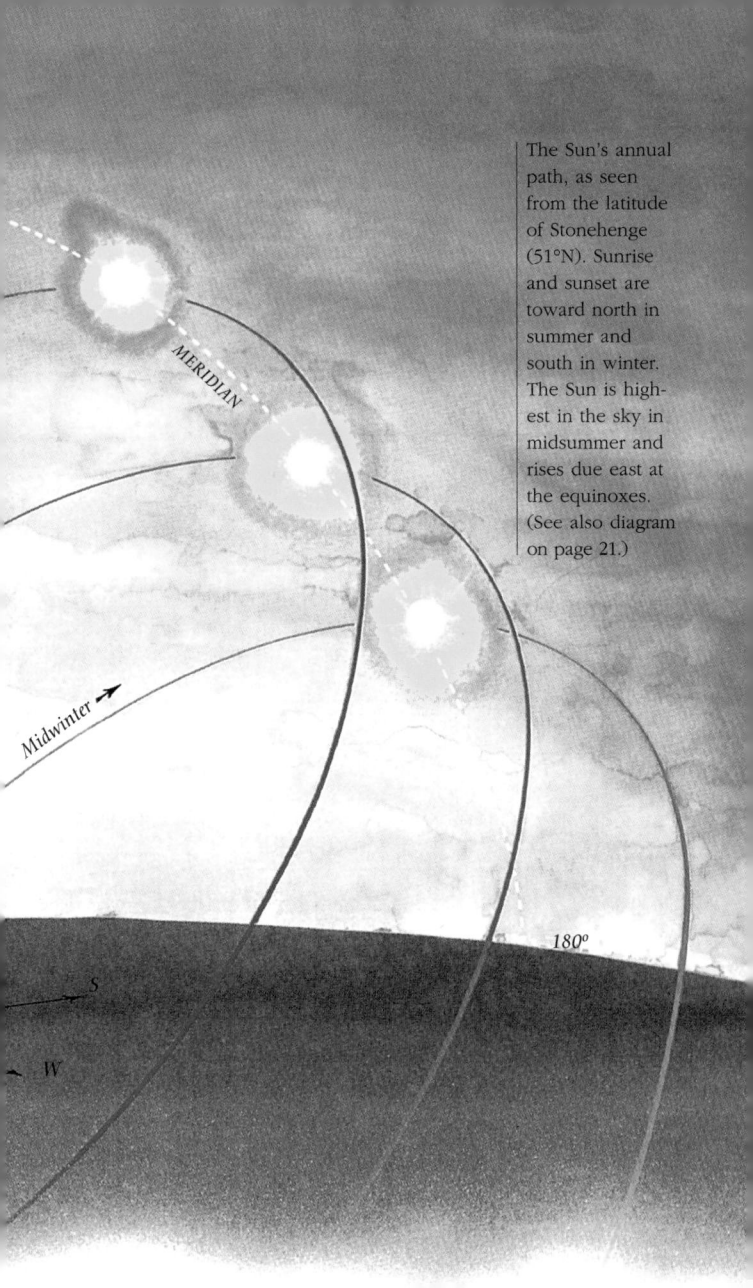

The Sun's annual path, as seen from the latitude of Stonehenge (51°N). Sunrise and sunset are toward north in summer and south in winter. The Sun is highest in the sky in midsummer and rises due east at the equinoxes. (See also diagram on page 21.)

MERIDIAN

Midwinter →

180°

S →

← W

THE MOON 1

The Moon 1

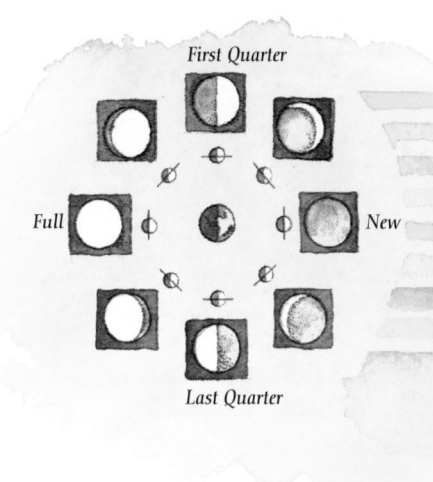

First Quarter

Full

New

Last Quarter

Over a 29.5-day cycle (slightly longer than the time the Moon takes to orbit the Earth), we see a changing proportion of the Moon's sunlit side, from New Moon, when the side facing Earth is all dark, to Full Moon, when it is all sunlit.

The Moon, a rocky body a quarter the size of our planet, is the Earth's only natural satellite. It orbits at an average distance of only 250,000 miles (400,000 km), and therefore seems large in the sky. Its pull on the Earth, together with that of the Sun, is responsible for the tides. In their turn, the gravitational forces of the Earth and Sun have slowed the Moon's rotation so that it now matches its orbital period, and only one hemisphere of the

Moon ever faces the Earth. By coincidence, from the Earth, the Moon appears to be the same size as the Sun, a fact that gives rise to eclipses that must have seemed astonishing centuries ago. It is little wonder that the ancients were interested in the Moon's complex movements. In fact, the Moon has a far more complicated cycle than the Sun, taking 18.6 years to complete.

Perhaps the most obvious feature of the Moon is its phases. Over a 29.5-day period the satellite waxes (grows) from a dark New Moon through crescent (literally "growing") to first quarter, half, "gibbous" (more than half) and then Full, before waning back to gibbous, half, decrescent and New. The 29.5-day cycle of phases is slightly longer than the Moon's orbital period (see page 28), as the Moon has to "catch up" with the apparent distance travelled by the Sun before each New Moon can occur.

The Moon has no light of its own, and shines only by reflecting sunlight

The Moon 1

The Moon 1

from its hemisphere that faces toward the Sun. Over the course of a lunar month, the Moon's orbit takes it from a position directly between us and the Sun, with light shining only on the side facing away from us – a New Moon – to a position directly opposite the Sun in the sky, so that we see the Earth-facing hemisphere fully illuminated. The crescent and decrescent Moon's dark regions sometimes glow faintly in sunlight reflected back from clouds and oceans, a phenomenon known as "earthlight".

From the Earth, the Moon's features, caused by impact craters and planes of solidified lava, appear to change every

Eclipses occur only when the Moon lies on the ecliptic, which happens twice in each orbit. In a solar eclipse, the Moon's shadow on the Earth is quite small, limiting the area on Earth from which the eclipse can be seen. But in a lunar eclipse, the Earth's shadow is much wider and thousands of miles long, so the eclipse can be seen for longer and over a larger area.

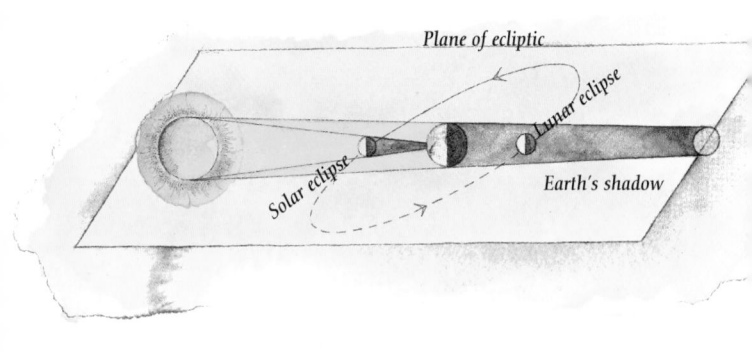

Plane of ecliptic

Solar eclipse

Lunar eclipse

Earth's shadow

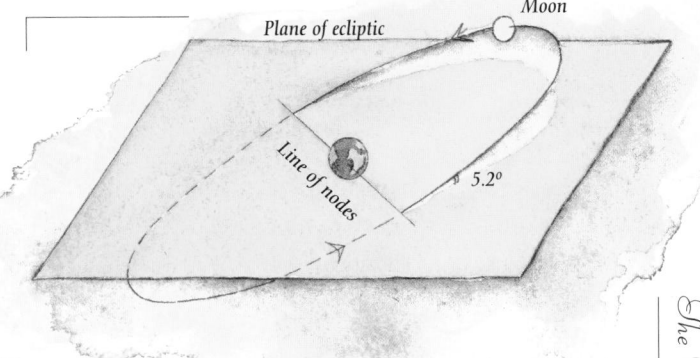

Moon

Plane of ecliptic

Line of nodes

5.2°

night as the angle of illumination changes. Shadows are strongest at first and last quarter, while the Full Moon can appear "washed out" by direct sunlight. The Moon's appearance alters so rapidly that lunar symbolism in many different cultures features the aspect of mutability.

As well as the total solar eclipse, there are less spectacular forms. Annular eclipses are where the Moon, at the outer reaches of its orbit, appears just too small to cover the Sun completely, displaying a bright ring. Partial eclipses are where the Moon or Sun is only partly obscured. The Moon itself can be eclipsed by the Earth (see diagram opposite).

The Moon's orbit is not a perfect circle but an ellipse. The inclination of the orbit at 5.2° to the plane of the ecliptic means that the Moon spends most of its orbit above or below the Earth-Sun plane. The points at which the lunar orbit crosses the ecliptic are the "nodes" (ascending and descending); the line through the Earth linking them is the "line of nodes".

THE MOON 2

The Moon orbits the Earth once every 27.33 days. If we could "freeze" the rest of the sky, the Moon's path through a single month would be fairly clear – a full circuit of the sky, with half its orbit below the ecliptic and half above, appearing at a maximum of 5.2° above or below the ecliptic. However, the Moon's path through the sky changes each month in a complex cycle that takes 18.6 years to complete.

Because the Earth is inclined at 23.5° to the ecliptic, the Moon shows seasonal variation similar to that of the Sun. This is clear when we remember that, for example, the Full Moon must always be on the opposite side of the sky to the Sun. In order to picture the Moon's movements, it often helps to concentrate only on the Full Moon, and to ignore the monthly rotation. Over a year, successive Full Moons move around the sky, always opposing the Sun. In Northern

The Moon 2

The gravitational forces of the Earth and Sun pull the Moon's orbit around in a circle once every 18.6 years. The nodes, where the Moon crosses the ecliptic, also move around with the orbit, an effect known as the "precession of the nodes".

midsummer, the Full Moon rises in the southeast and sets in the southwest, and hangs low throughout the night. In midwinter, when it passes high overhead, it rises in the northeast and sets in the northwest. Over a longer period, these extreme southerly and northerly rising points move as well: at one time, the extremes of the Moon's motion are close together, and 9.3 years later they are at their widest separation.

This second element of the cycle is due to the fact that the Moon's elliptical orbit also shifts, revolving around the Earth once every 18.6 years. When the Full Moon is at the outermost point on the shifting elliptical orbit, it is also fur-

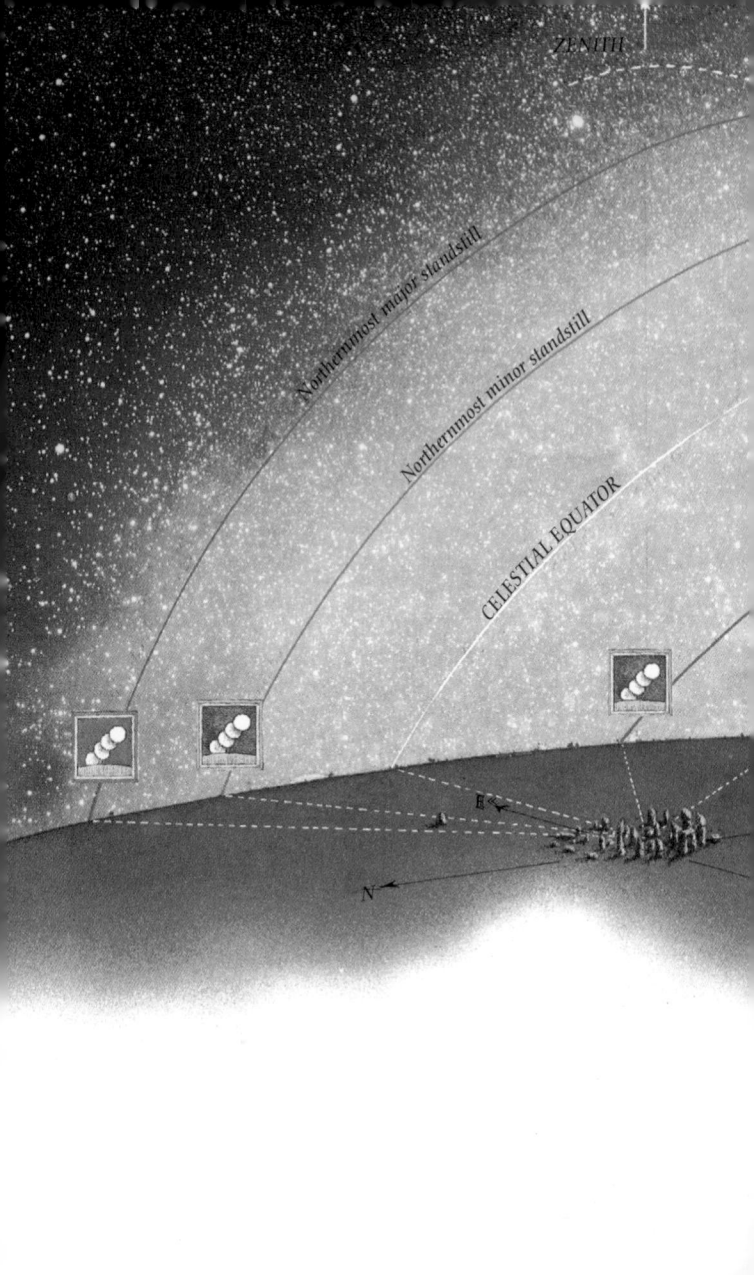

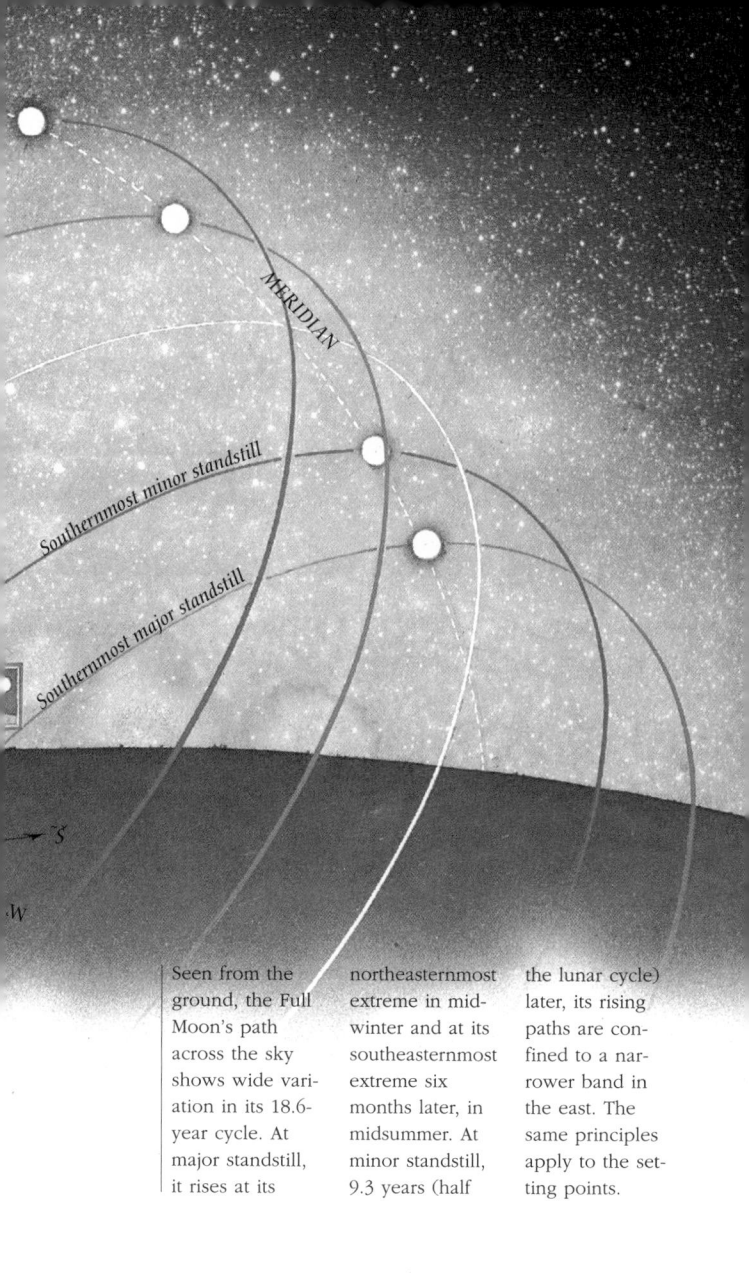

MERIDIAN

Southernmost minor standstill

Southernmost major standstill

→ S

·W

Seen from the ground, the Full Moon's path across the sky shows wide variation in its 18.6-year cycle. At major standstill, it rises at its northeasternmost extreme in midwinter and at its southeasternmost extreme six months later, in midsummer. At minor standstill, 9.3 years (half the lunar cycle) later, its rising paths are confined to a narrower band in the east. The same principles apply to the setting points.

The Moon 2

thest from the ecliptic (see illustration, pages 28–29). A "major standstill" occurs when the Moon shows the widest annual range of motion in its 18.6-year cycle. It is called a standstill for the same reason that the solstices are so called ("solstice" means "sun standstill"): the Full Moon's cycle of movements apparently halts before reversing. At the other end of the cycle lies the "minor standstill", when the Full Moon shows its narrowest annual range of motion (see pages 30–31).

As the Moon's orbit revolves, the line of nodes where it crosses the ecliptic turns as well. Since eclipses can only occur when the New or Full Moon lies on the line of nodes, the motion of this line creates an 18.6-year eclipse cycle. However, because 18.6 years does not equate to a complete number of lunar months, the Moon will not return to the same point in its orbit until it has gone through three such cycles. Hence, there is a longer cycle of just under 56 years.

THE PLANETS

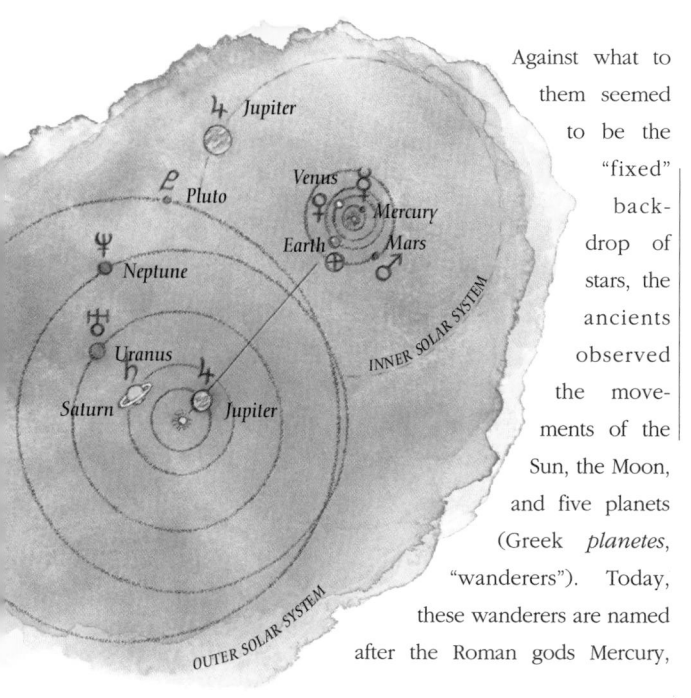

The Planets

Against what to them seemed to be the "fixed" backdrop of stars, the ancients observed the movements of the Sun, the Moon, and five planets (Greek *planetes*, "wanderers"). Today, these wanderers are named after the Roman gods Mercury,

THE OUTER PLANETS (BELOW) AND THE INNER SOLAR SYSTEM

Jupiter: The largest planet. Orbits the Sun in 11.86 years.
Saturn: Orbits in 29.46 years. Best known for its spectacular rings.
Uranus: Tilted at 98°, it "rolls" around its orbit in 84 years.
Neptune: A small gas giant, similar to Uranus, orbiting in 164.8 years.
Pluto: Its elliptical 248.6-year orbit brings it closer than Neptune to the Earth for a brief period.

The Planets

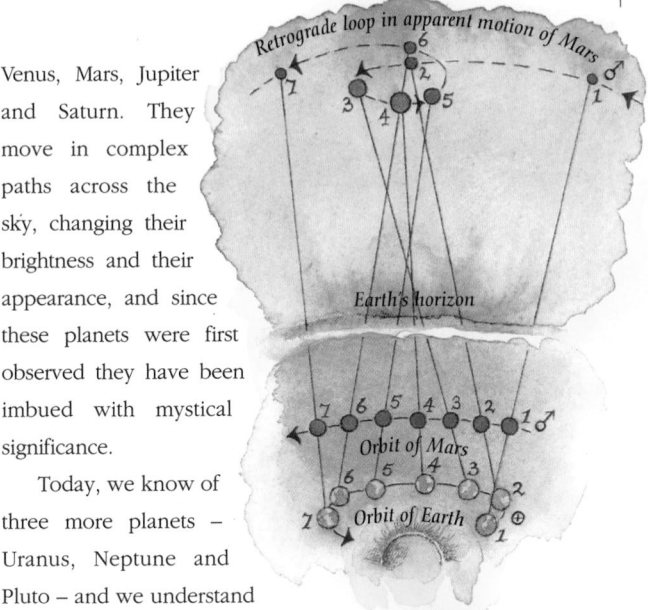

Venus, Mars, Jupiter and Saturn. They move in complex paths across the sky, changing their brightness and their appearance, and since these planets were first observed they have been imbued with mystical significance.

Today, we know of three more planets – Uranus, Neptune and Pluto – and we understand that all eight are fellow members of our Solar System. The inner Solar System contains relatively small, rocky worlds: Mercury, Venus, the Earth and Mars. Beyond the asteroid belt (a ring of rocky debris from the birth of the Solar System) that lies between Mars and Jupiter lie four "gas giant" planets: Jupiter, Saturn, Uranus and Neptune. Pluto, a tiny, rocky world beyond the gas giants, is

The apparent retrograde (backward) motion of some planets is caused by the changing line of sight as the Earth passes close. For example, Mars appears to complete a slow loop against the stars in the sky.

Strikingly, Pluto orbits at 17° to the ecliptic. Halley's Comet, the orbital period of which is 76 years, has a sharply inclined elliptical orbit, a pattern common to most comets – small chunks of rock and ice vaporized by the Sun's heat.

probably the largest member of another belt of debris that rings the Solar System.

Mercury and Venus, with orbits smaller than Earth's, are often called "inferior" (lower) planets. Their paths always lie close to the Sun, so they are usually seen near sunset or sunrise. Venus, the closest planet to the Earth, has a highly reflective atmosphere which makes it the third brightest object in the sky. It is known as both the Morning

The Planets

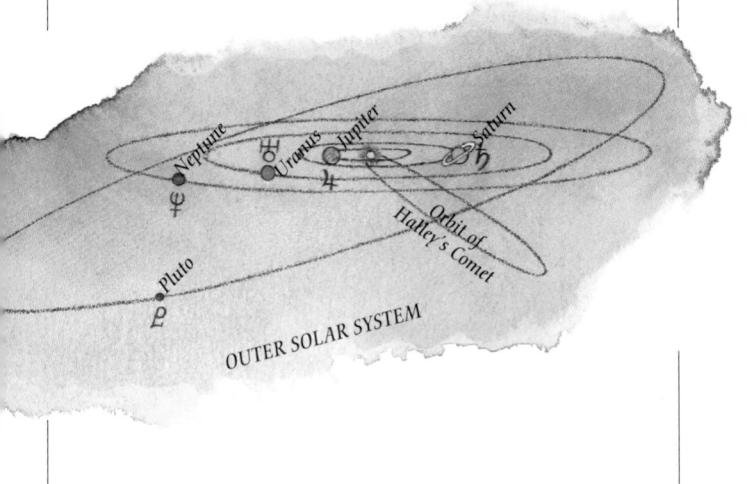

OUTER SOLAR SYSTEM

The Planets

and the Evening Star from its appearances near to the rising and setting Sun. Through binoculars we can see that the inferior planets show phases similar to the Moon's. They also pass across the face of the Sun, although their apparently smaller size leads to transits (literally "crossings") rather than eclipses.

When an inferior planet and the Earth line up on opposite sides of the Sun, this is known as a "superior conjunction"; and when an inferior planet lies at its closest to us, aligned between the Sun and the Earth, this is an "inferior conjunction". The "superior" (upper) planets can also come into conjunction, both with each other – when they appear to line up in the sky – and with the Sun, which happens when the Earth, the Sun and a superior planet align in that order. An "opposition" occurs when a superior planet is at its closest to the Earth – in line with the Earth and the Sun but, unlike in an inferior conjunction, on the opposite side of the Earth from the Sun.

THE STARS

The Stars

A star's position on the Celestial Sphere is given as declination in degrees above or below the celestial equator, and Right Ascension in time from the First Point of Aries, where the ecliptic crosses the celestial equator.

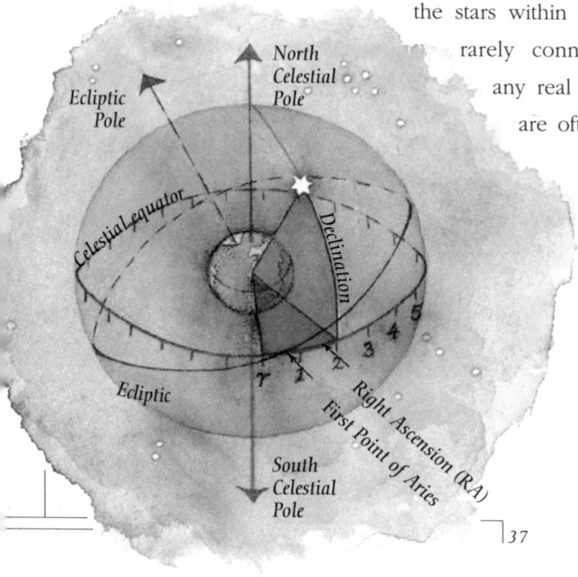

Unlike the Sun, Moon and planets, the stars do not change position noticeably over short periods, except for the apparent motions caused by the Earth's rotation. Their locations on the Celestial Sphere (see page 19) can be plotted with remarkable accuracy, and consequently constellations have been used for centuries as stationary markers to chart the motions of other celestial bodies. Cultures all over the world have identified constellations based upon mythology. They are, in reality, line-of-sight effects – the stars within them are rarely connected in any real way and are often many

Ecliptic Pole

North Celestial Pole

Celestial Equator

Declination

Ecliptic

Right Ascension (RA)

First Point of Aries

South Celestial Pole

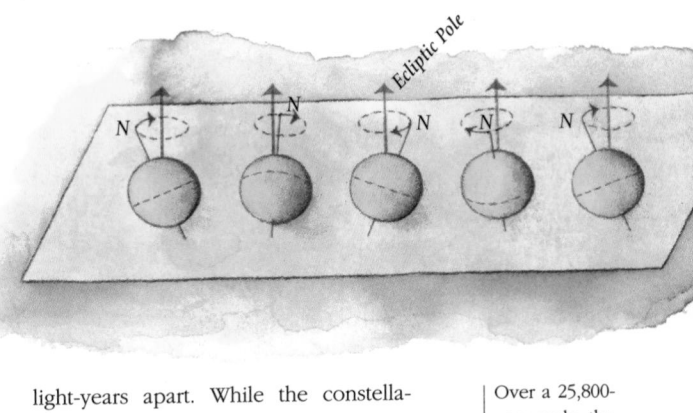

light-years apart. While the constellations seem to be static, in fact every star moves very slowly. In a few thousand years, most of the patterns now known will have gone. However, in the short term, stars do not seem to change relative positions.

The coordinates of stars are recorded in terms of "declination" (the angle from 0 to 90° above or below the celestial equator) and "Right Ascension" (RA). RA is the period between the moment when the First Point of Aries (the first house of the zodiac) crosses the meridian and the moment when the object of interest crosses the same point – giving

Over a 25,800-year cycle, the combined gravitational pull of the Moon and the Sun causes the planet to wobble like a slow spinning-top before it falls over.

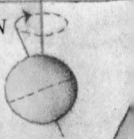

a reading in hours, minutes and seconds, up to 24 hours. The North Celestial Pole, with a declination of +90°, is marked by the star Polaris, which lies within half a degree of it: the South Celestial Pole (-90°) is marked only by the faint star Sigma Octantis.

The paths of the stars are affected by the Earth's motions, but over a much

At 35° South, the Southern Pole Star is 35° above the horizon, the Southern Cross is circumpolar (never setting), and Gamma Centauri dips out of view.

The Stars

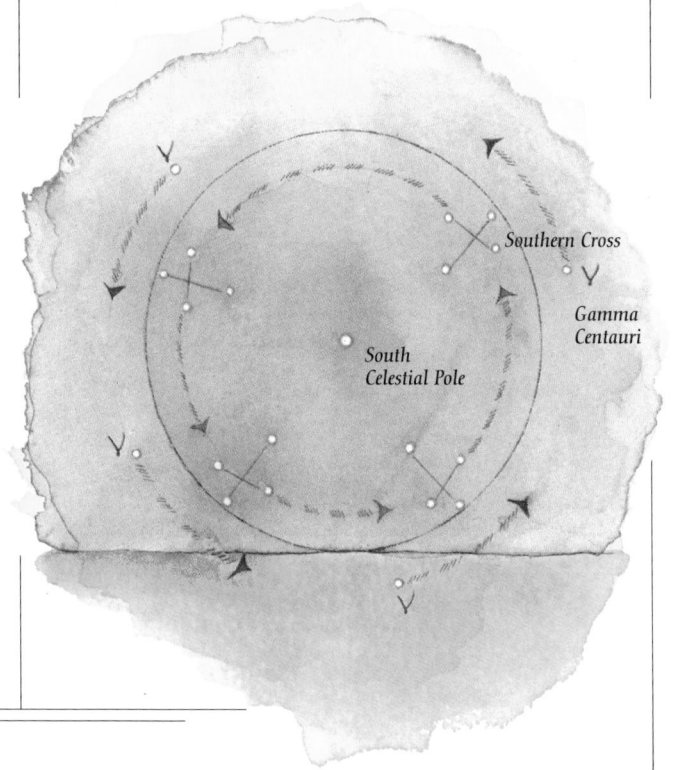

Southern Cross

Gamma Centauri

South Celestial Pole

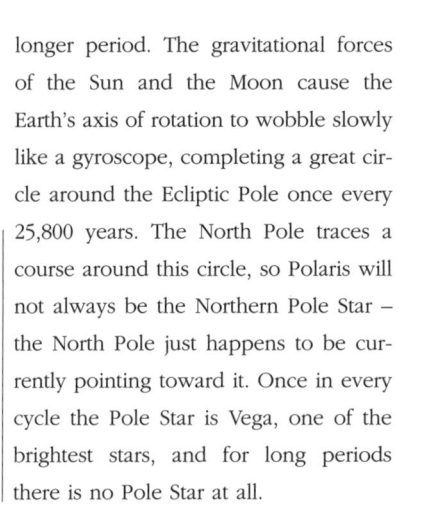

The Stars

longer period. The gravitational forces of the Sun and the Moon cause the Earth's axis of rotation to wobble slowly like a gyroscope, completing a great circle around the Ecliptic Pole once every 25,800 years. The North Pole traces a course around this circle, so Polaris will not always be the Northern Pole Star – the North Pole just happens to be currently pointing toward it. Once in every cycle the Pole Star is Vega, one of the brightest stars, and for long periods there is no Pole Star at all.

Because the Pole is constantly changing, the paths of the stars change slowly too. This effect, termed precession of the equinoxes, entails complicated calculations to correct the RA and declination of objects – astronomers have to update their atlases at least every 50 years. Only the apparent motions of Solar System objects are unaffected: the Earth's inclination to the ecliptic remains a constant 23.5°, so the ecliptic is the one fixed line in the sky.

THE ZODIAC

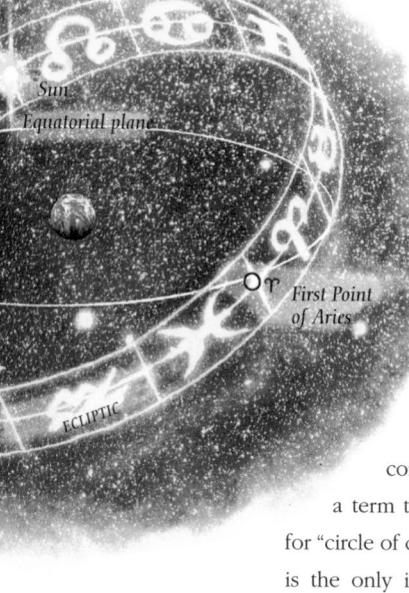

Sun

Equatorial plane

☿ *First Point of Aries*

ECLIPTIC

The Sun, Moon and planets appear to us to move against the background stars in a very narrow band of sky, because they all orbit on or close to the ecliptic plane. The significant stars around this band are divided into the twelve constellations of the "zodiac", a term that comes from the Greek for "circle of creatures" (Libra, the scales, is the only inanimate zodiacal image). Astronomically, there is also a thirteenth zodiacal constellation, Ophiuchus the healer, into which the planets, as well as the Sun and Moon, also venture; however, astrologers usually ignore it (see pages 192–195).

Modern astrology is based on the writings of the Greek astronomer-

The zodiac can be imagined as a band encircling the Earth with the ecliptic along its centre. At 0° Aries, the Sun's path on the ecliptic crosses the equator heading north.

The Zodiac

The Zodiac

astrologer Ptolemy of Alexandria (2nd century BCE), but its roots go back much further (see pages 82–87). Simplifying the astronomical zodiac, astrology divides the zodiac' band into twelve equal sectors called "houses" of 30° each.

When the zodiac was first calculated, the Sun crossed the equator heading northward at the boundary of the houses of Aries and Pisces – this moment marked the beginning of Northern spring, and is termed the Origin, or First Point, of Aries. Over thousands of years, the precession of the equinoxes has moved this point into the constellation Pisces. This has led to a major departure of method between Indian astrology, which has stayed with a fixed zodiac pegged to the actual constellations, and the European/Islamic tradition, which moves its zodiac with the spring equinox point. Astrologers suggest that both zodiacs, moving and fixed, have their own relative orders of symbolic truth.

This illustration, based on a drawing by Sebastian Verney, shows the zodiac constellations on the ecliptic between Virgo and Pisces at mid-European latitudes. Virgo is in the ascendant, rising in the east, with the autumnal equinox point just on the eastern horizon.

ZENITH

MERIDIAN

ECLIPTIC

CELESTIAL EQUATOR

E

S

W

♈ First Point of Aries

This illustration of *c.*950CE refers to the Book of Revelation, 12:1–18. Apocalyptic scenes are played out against a background of Earth, Heaven and Hell. The woman "clothed in the Sun" bears a son, whom the dragon tries to devour. God snatches the child to safety. Angels fight the dragon, banishing it to Earth. Here, the dragon chases the woman and tries to drown her.

THE GRAND DESIGN

Every culture possesses a strong impulse to map the cosmos. The tendency in early civilizations was to imagine a vertical hierarchy, with mysterious realms above and below the visible. In the West, with the development of scientific knowledge, a spherical conception evolved, reflected in the way in which we imagine the Celestial Sphere (see pages 19–20 and pages 57–58). The power of these ancient images comes from their evocation of subjective human experience. Their poetic vision strikes our imagination, if not our reason, with the force of a profound truth.

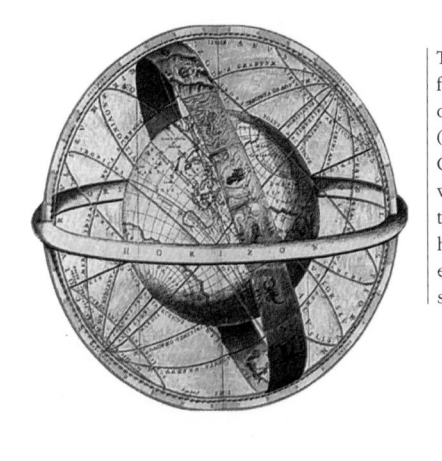

This engraving from an atlas of the heavens (1660) shows the Celestial Sphere with the band of the zodiac and a horizon projected onto the sphere.

THE EARTH AND SKY

The primordial phenomena of Earth and Sky form the basis of our understanding of space and, as a consequence, time and order. The Earth on which we live is dominated in every direction – before, behind and at either side of us – by the overarching vault of the heavens above: this is how we know space. We experience the alternation of day and night as the Sun rises and sets around us: this is how we know time.

The root myths of all cultures include cosmogony, a description of the origin of the cosmos, in which the Earth-Sky pair feature among the first elements. In ancient times, the priests of Heliopolis, the City of the Sun in northern Egypt, taught that in the midst of the formless primordial ocean (Nun), slumbering in a lotus bud, lay Atum, whose name means "the all" or "the complete". Eventually, by the power of will, he drew himself out of non-being

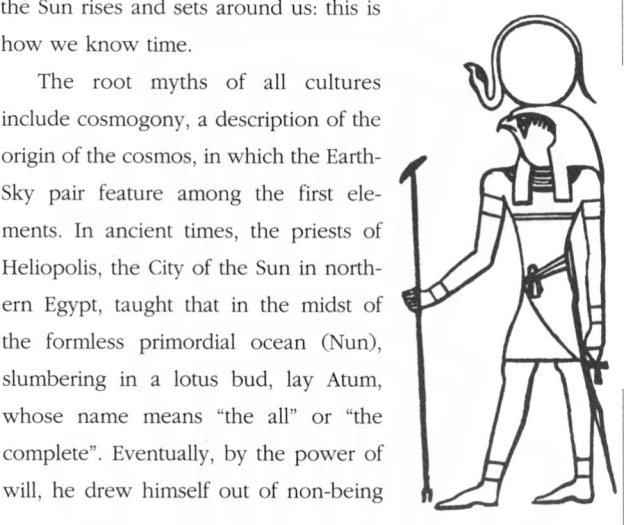

The Earth and Sky

This papyrus dating from *c*.1000BCE shows the Egyptian sky-goddess Nut arched above her brother and lover Geb, the Earth, who lies on the Earth trying to reach her. Shu is shown between them, forcibly separating the Earth from the Sky.

and manifested himself as the Sun god. This supreme divinity bore from himself twins – Shu, a boy, and Tefnut, a girl – who in turn produced Geb, the god of Earth, and Nut, the goddess of Sky.

These two also lay together and Shu, the god of air and space, forced his way between them to prise them apart. Nut was thrust upward, and there she stays stretched out above the Earth for eternity in her classic posture, supported by her arms and legs – the four pillars of the firmament. The stars lie along her

belly, and every day from rising to set-ting, the boat of the Sun god passes along her body. Below, Geb wails and tries to lift himself to his beloved, and the contours of his agonized body form mountain chains on Earth.

Nut was pregnant by Geb but, in one account, the Sun god henceforth decreed that she could not bear a child in any month of the year. But the god Thoth played checkers with the Moon and won a seventy-second part of the Moon's light, with which he created five new days that did not belong to the 360-day Egyptian calendar. Thus Nut could bear her five children: Osiris, Isis, Horus Seth and Nephthys. (Horus is usually said to be the son of Isis and Osiris.)

The appearance of the Sky goddess in a mythic description of intercalation (adding days to the calendar to keep it in line with the solar year) alerts us to an important principle: cosmogonic myths go beyond Earth-Sky as a spatial entity and refer to time and time-keeping,

The Earth and Sky

which are themselves derived from celestial motions and are of the greatest importance for both the practical and ritual regulation of ancient society.

Hesiod's *Theogony* (*c*.750BCE) is the earliest Greek account of creation. It has significant parallels with the Egyptian story, which the Greeks themselves acknowledged. Thus, in the Greek myth we also find the union of Earth and Sky being forcibly prevented. After dark Chaos, there came fertile Gaia, the Earth, who produced Uranos, the Sky, crowned with stars and equal in magnificence to Gaia herself. Uranos covered the Earth and mated with her. However, he forced their offspring – three 100-handed giants, three Cyclopes, and twelve Titans – back into Gaia's womb. In revenge, Gaia made a sickle for her last-born, the Titan Cronos, who castrated his father when he next came to mate with Gaia. The genitals fell into the sea with a splash of white foam from which sprang the goddess Aphrodite.

The separation of the two realms, often linked to the misdemeanours of humankind, is a recurrent theme in mythology all around the world. The heavenly and terrestrial worlds are also commonly said to be linked by a tree or ladder, as in the Western children's tale of Jack and the Beanstalk. The notion of the sky as home of the gods is equally widespread, and explains the sacred significance of high-soaring birds such as the eagle. For the Aztecs, the eagle was the power of the rising Sun, enemy of darkness, embodied by a serpent. Many Native American peoples revere the eagle-like Thunderbird, the mighty messenger between the Earth and the Sky, whose wingbeats are heard as thunder.

The Earth and Sky

The Egyptian Sky goddess Nut stretches over the Earth with the stars, Sun and Moon passing through her body. The illustration is based on a painting inside the lid of a coffin dating from the 11th–8th centuries BCE.

HEAVENLY ORDER

Heavenly Order

The birth of the universe as recounted in the Mayan creation epic, the *Popol Vuh* (a collection of texts collated in the 16th century), is mysterious and evocative: "Whatever might be is simply not there: only murmurs, ripples, in the dark, in the night. Only the Maker, Modeller

This 15th-century woodcut from an English "shepherd's calendar" shows the heavenly spheres and the Sun (on the right) and Moon (on the left).

alone, Sovereign Plumed Serpent, the Bearers, Begetters are in the water, a glittering light."

Featuring in this and many other cosmogonies is the archetypal figure of the maker-god who establishes the cosmic design and creates humankind as a key element in the great work. Without humans, who would there be to know, praise and serve the god?

The image of the maker of the world as a heavenly smith is a recurrent motif in mythology. In Nordic and Anglo-Saxon myth we find a shadowy figure, often lame, variously known as Waldere, Volund or Wayland the Smith. His Greek counterpart is Hephaistos (Vulcan to the Romans), god of fire and forge, who was born lame and weak but was a master craftsman and builder. From his anvil, with twenty bellows working spontaneously at his bidding, he built all the palaces on Mount Olympus.

The heavenly smith performs his most celebrated work in the *Timaeus*,

Heavenly Order

Heavenly Order

the cosmological myth drawn from archaic sources by Plato (*c.*428–348BCE), which dominated Greek, Islamic and European science and philosophy for two millennia.

Plato's maker-god, the Demiurge, created the universe in the image of the most perfect form, a single rotating sphere – hence the word "universe" itself (Latin *universum*, "whole", from *uni-vertere*, "to make one turn"). The perfect being of the heavens is that which turns in one plane on its axis – in other words, the Celestial Sphere (see page 19). The sphere was created by a harmonious arrangement of the four elements: fire, earth, air and water. Fire gave it visibility and earth made it tangible; and these two were bonded together to form an indissoluble unity by the mediation of water and air.

The Demiurge wrapped the spherical being in the "world soul", the *anima mundi* as it was later termed. This soul is compounded of existence, sameness

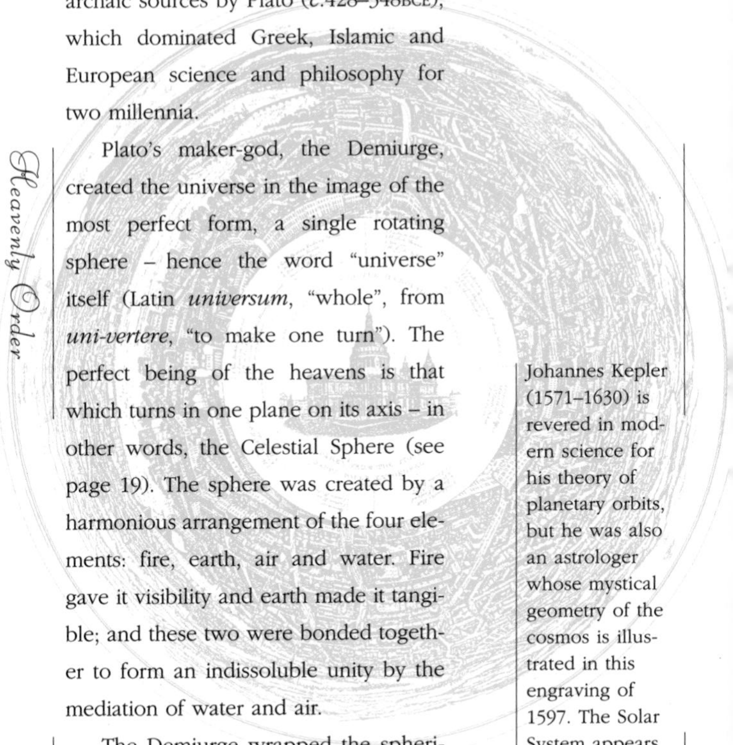

Johannes Kepler (1571–1630) is revered in modern science for his theory of planetary orbits, but he was also an astrologer whose mystical geometry of the cosmos is illustrated in this engraving of 1597. The Solar System appears here as a series of planetary spheres and "Platonic" solids.

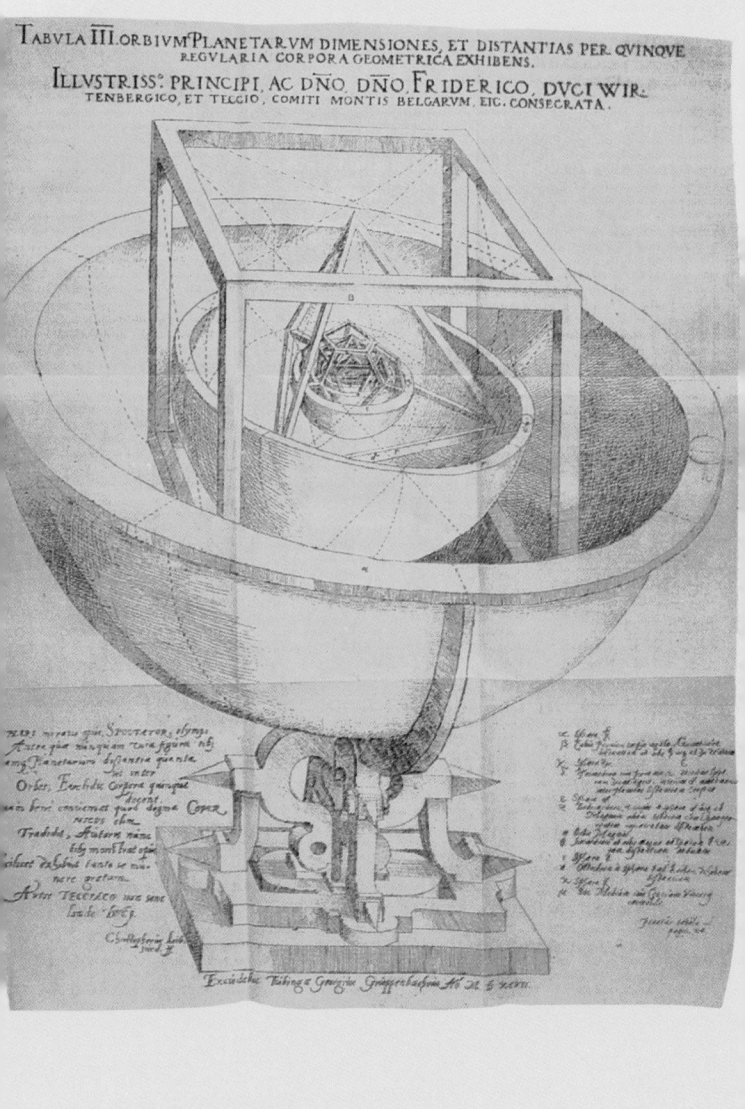

TABVLA III. ORBIVM PLANETARVM DIMENSIONES, ET DISTANTIAS PER QVINQVE
REGVLARIA CORPORA GEOMETRICA EXHIBENS.

ILLVSTRISS: PRINCIPI, AC DÑO. DÑO. FRIDERICO, DVCI WIR-
TENBERGICO, ET TECGIO, COMITI MONTIS BELGARVM, EIC. CONSECRATA.

Excudebat Tubingæ Georgius Gruppenbachius Aõ M D XCVI.

Heavenly Order

and difference – existence is the condition for anything to be, and all things are distinguished and known as themselves on account of their similarity to, or difference from, other existing things.

The Demiurge created the equator and the ecliptic circles of the Celestial Sphere by splitting the fabric of the world soul into two strips which he curved into circular bands. He laid one across the other at a slant of 23°. The ecliptic circle comprised the seven circuits of the planets which moved around the Earth at their centre. The remaining soul material became the souls of humankind, sown in the heavens before incarnation into the brutal realm below.

For Plato, the neo-Platonists and the Hermetic astrologers who came later, there was no higher purpose in our study of the heavens than the recovery of our primal perception of the goodness and greatness of the universe, from which our substance and our intelligence are derived.

THE COSMOS,
HUMAN AND DIVINE

Plato's pupil, Aristotle (384–322BCE), takes us from a system of thought that allies naturally with poetry and myth, into a realm of rationality, logic and science. However, at root these two fathers of philosophy share a common concept of the construction of the cosmos. This cosmological legacy binds together many distinct and seemingly unrelated expressions of religion, magic, science, philosophy and astrology over nearly two millennia in the Greco-Roman, Persian, Islamic and later European cultures.

Aristotle effectively translated the groundplan of Plato's cosmos into a scientific conception. Like his teacher, Aristotle saw the Celestial Sphere as the perfect creation of the creator-god, the Zeus of abstract philosophical principle rather than the temperamental god of the common people and their myths. The sphere is incorruptible and eternal,

The Cosmos, Human and Divine

The Cosmos, Human and Divine

its stars are divine fire, its motion the first of all movements. All possible motions, whether in the heavens or on the Earth, are ultimately derived from the rotation of the Celestial Sphere around its own axis.

The coming-into-being and passing away of all things in creation take their cue from the primary

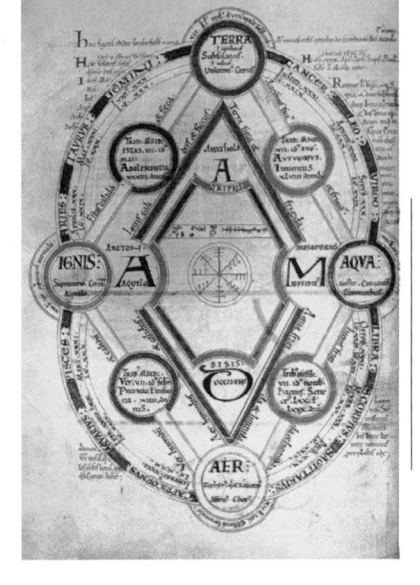

VIRTUES OF THE HEAVENS OF PARADISE, ACCORDING TO DANTE

	Heaven	Souls Encountered	Discipline or Virtue
1st	Moon	Inconstant in their vows	Fortitude required
2nd	Mercury	Active and ambitious	Justice required
3rd	Venus	Lovers	Temperance required
4th	Sun	Theologians, teachers, historians	Prudence
5th	Mars	Warriors	Fortitude
6th	Jupiter	The just	Justice
7th	Saturn	Contemplatives	Temperance
8th	Stars	Virgin Mary, saints and Adam	Faith, Hope and Love
9th	Primum Mobile		
10th	Empyrean	The Blessed; St Bernard	

VI
ORDINAMENTO DEL PARADISO

The Cosmos, Human and Divine

cosmic cycle – the anticlockwise movement of the Sun along the ecliptic. This is similar to Plato's formulation, whereby distinctions between things are caused by the motion of the planets along the ecliptic.

In its final form, Aristotle's cosmology and physics, coupled with the sophisticated theory of solar, lunar and planetary motions

OPPOSITE, ABOVE
This English cosmological diagram of *c.*1110CE shows the four elements linked by the zodiacal constellations.

ABOVE The universe according to Dante. At the bottom, the four elements are shown with the Moon. The planets then come in

Ptolemaic order, with the Sun after Venus. After Saturn, toward the top of the picture, the stars of the zodiac are represented.

The Cosmos, Human and Divine

of his contemporary, Eudoxus (*c*.400–350BCE), provided a complete basis for an understanding of the physical universe. The final phase in the construction of this model was supplied by Ptolemy of Alexandria (2nd century CE), and it is his name (Ptolemaic) that is given to the long-enduring Earth-centred construction of the universe known to later Islamic culture and to medieval Europe.

The most sustained and beautiful medieval account of the cosmos is Dante's *The Divine Comedy* (*c*.1310–*c*.1320), a metaphor of the path of the soul through pagan and Christian mysteries, unfolded in the universe of Plato, Aristotle and Ptolemy. In the third book, *Paradise*, the ascent into Paradise occurs through successive planetary spheres (see illustration, page 59). Dante's description of this ascent uses established astrological symbolism to create a hierarchic catalogue accommodating the souls of the illustrious dead, angels and events in history (see table, page 58).

CELESTIAL METAPHORS

In attempting to conceptualize the cosmos, many cultures have drawn upon metaphors derived from terrestrial observation, and, in the process, have created some profound and poetic images of the universe.

A classic example is the widespread metaphor for the polar axis, the World Tree. The pagan religion of northern Europe, preserved for us in the Scandinavian prose tales known as the *Edda*s, tells of a giant ash tree, Yggdrasil, which comprises the whole world. One of its three roots plunges into Niflhel, the underworld; the second root is in the icy land of the giants; and the third root reaches up to heaven. In the Mayan culture of Central

In this Greco-Roman period Egyptian coffin painting, the *ba*, or soul (right), receives water from the goddess in the Tree of Life (left).

Celestial Metaphors

America, whose mythology has only recently been revealed as deeply sky-oriented, the World Tree is a central conception. The name of the Tree, Wakah-chan, means "raised-up sky": in Maya myth, First Father raised it at the beginning of creation in order to divide Sky from Earth.

In some cultures, the celestial metaphor that orientates the universe is more elaborate, although still based on the idea of a mysterious centre to creation. The Navajo people of Arizona tell the following story of creation. In the time of great darkness, Sky Father descended and Earth Mother rose to meet him, and on the top of the mountain on which their union had occurred, the ancestors of humankind found a small turquoise figurine. This became the immortal goddess Estsatleh, meaning "she who rejuvenates

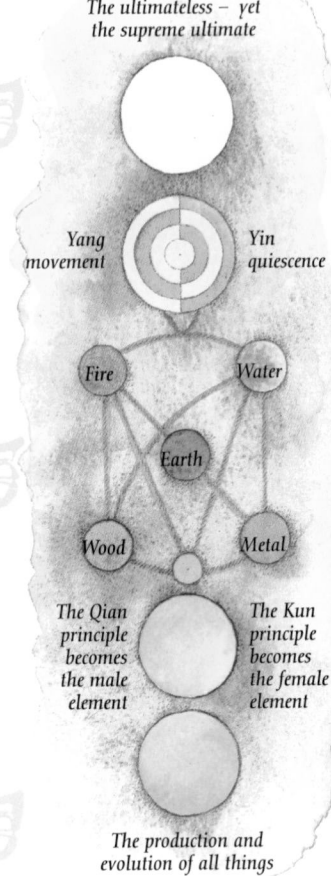

The ultimateless – yet the supreme ultimate

Yang movement

Yin quiescence

Fire

Water

Earth

Wood

Metal

The Qian principle becomes the male element

The Kun principle becomes the female element

The production and evolution of all things

Celestial Metaphors

herself": after passing through maturity and becoming a crone, she rose again as a girl. Four daughters were born from parts of her body, and a fifth was born from her spirit. The Sun came out of the turquoise beads on her right breast and the Moon emerged from the white shells on her left breast.

In this story, we see one of the many variations on a recurring cosmological motif – a mysterious place where Sky and Earth join and the lights of the sky are born. This is the place where the four directions and the four seasons are demarcated, with the centre as the mysterious "fifth place". Estsatleh is complete unto herself, turning around her own centre in perpetual rejuvenation. For us the movement of this god is the rotation of the Celestial Sphere (see pages 19–20 and 57–58), and the mysterious place where the Navajo ancestors found the turquoise figurine represents the polar axis of the heavens, around which the whole of creation revolves.

A diagram of the "Supreme Ultimate" of Chinese philosophy. After the inexpressible ultimate of being and non-being, the opposing but interdependent energies of Yin (originally light) and Yang (originally dark) interact to produce all phenomena in the universe.

Celestial Metaphors

The mysterious place at the cosmic centre is sometimes represented as a stone. At Delphi, site of the most famous oracle in ancient Greece, there still stands the archaic rounded block of stone known as the *omphalos*, or "navel". The Greeks revered the site as the very centre of the Earth, the navel of the great goddess Gaia, whose shrine this originally was.

The same symbolism springs to life with surprising consistency in several myths that on the surface appear to have little in common. The shamans of the scattered Finno-Ugric tribes that once stretched from Lapland to Siberia had as a central symbol a mysterious talisman called Sampo, forged by the smith Ilmarinen. But could this character be another guise of Wayland the Smith and Plato's Demiurge (see pages 53–54), forgers of the heavens? It has been fairly conclusively established that "many-coloured" Sampo is a representation of the star-spangled heavenly vault.

PRECESSION AND
THE GREAT AGES

The stars we see in the night sky have
all shifted their position in the frame-
work of the celestial equator and the
ecliptic since observations were made
by the early Greek astronomers. This
shift has been caused by a phenomenon
termed precession – the wobble of the
Earth's axis, in a movement akin to that
of a gyroscope, forming a complete
cycle over 25,800 years (see page 40).
The first scientific description of this
cycle is credited to Hipparchus (2nd
century BCE), but earlier cultures incorporated aspects of the phenomenon
into myth, especially the precession of
the equinoctial and solsticial points,
which appear to move backward against
the fixed stars by about 1° every 72 years.

How far do the myths of various cultures reflect these long-term celestial
changes? While mainstream academic
opinion resolutely avoids making an

Precession and the Great Ages

A representation of Aquarius pouring his water into the river of life. Owing to precession, the vernal (spring) equinox point will pass into Aquarius during the next millennium.

upward gaze, there is a minority view that the heavens are the significant determinant of both the form and substance of major myth-complexes. In the most comprehensive study along these lines, *Hamlet's Mill: Myth and the Frame of Time* (1969), Giorgio de Santillana and Hertha von Drechend claim that the geography that is revealed in myth – the cosmos that forms the "world" of archaic imagination – lies in the heavens.

Roads, rivers and oceans in many myths often refer to parts of the sky, especially to the path of the Milky Way. Similarly, poles and trees refer to the celestial axis.

Perhaps the most challenging possibility of this interpretation is the claim that precession has been assimilated into myth, so that, for example, the movement of the polar axis becomes the fall of the World Tree or the expulsion from Eden. It is believed that the cataclysm at the end of each age reflects the displacement of successive constellations as they lose their hold on the equinoxes and solstices.

In recent decades, precession has taken root in the popular imagination of the "New Age" of Aquarius. However, the assumption that we are at the dawning of this Great Age owes very little to observation of the sky. Since *c.*100BCE, the equinox point has been making its way slowly through the constellation Pisces and is only now beginning its progress through the second fish of the Pisces pair: it will not reach the same

Precession and the Great Ages

degree of longitude as the star Beta Piscium at the head of this fish until 2813CE. Even stretching the case, we would barely brush the edge of Aquarius before 2300CE.

Precession has been potently interpreted in our era by the psychologist Carl Jung (1875–1961): "The course of our religious history as well as an essential part of our psychic development could have been predicted, both as regards time and content, from the precession of the equinoxes through the constellation of Pisces." Thus, Christ, the "First Fish", was born on the crossover from Aries to Pisces, while the rise of science and secular rational-

OPPOSITE This fish in a fresco in Rome represents Christ, whose birth coincided with the spring equinox point entering the first fish of Pisces.

ism reflects the second fish. The political theorist and atheist Karl Marx was born in 1818 – within a year after the exact period (the ecliptic contact) in which the spring equinox point had the same degree of longitude as the first star in the second fish (Omega Piscium).

Precession and the Great Ages

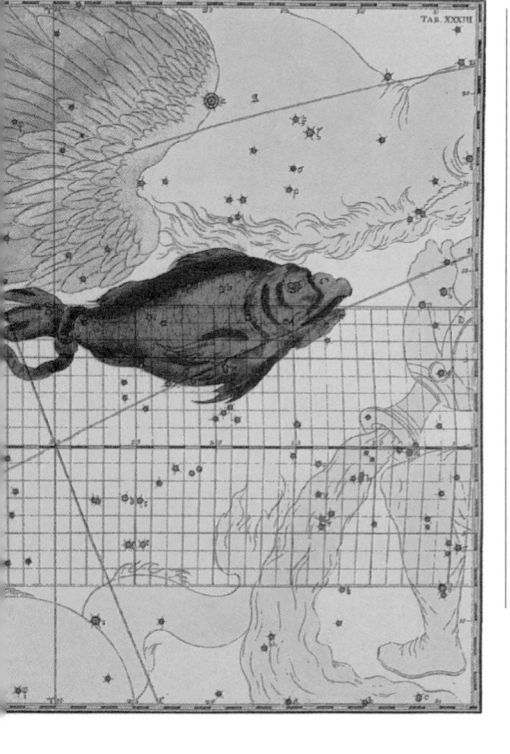

An 18th-century English engraving of the constellation Pisces. The first fish in the Age of Pisces is on the left. The birth of Christ occurred at a point between the alpha star in the knot of the cords and omicron further up the cord attached to the first fish. Karl Marx was born near the omega star, which is just on the knot at the tail of the second fish.

69

This Flemish tapestry, hanging in Toledo Cathedral, dates from the 15th century and shows zodiacal constellations alongside other star-groups which are pictorially represented. The floral background reinforces the idea of a heavenly realm that interpenetrates the terrestrial.

CORRESPONDENCES

For ancient cultures, the celestial cycles provided a pattern for the world below. Even where the heavens might be denied a governing role, as in the monotheistic religions, the stars were nevertheless considered a supreme expression of divine will and order.

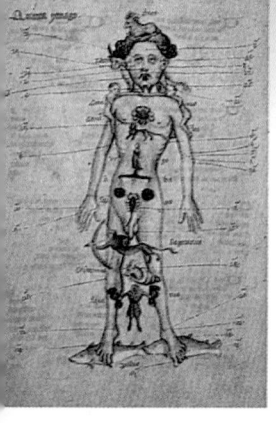

A late medieval image of Zodiacal Man, a concept that saw the idealized human form as a microcosm of the heavens.

Greek and Roman philosophers identified the microcosm, the "little world" below, with the macrocosm, the "great world" above. The Hermetic axiom, "As above, so below", attributed to the ancient mythical sage Hermes Trismegistos, expresses this notion of correspondences, which retains its vitality – most notably in astrology – to the present day.

THE STARS, ORACLES AND FATE

In the most sophisticated astrologies, the relationship between character, fate and the heavens is both subtle and natural, reflecting not a disruptive interference from external forces but a complex and universal harmony within the universe. Since the 18th century Enlightenment, science has gained an ever-advancing dominion over the material universe. Yet many people today believe that the ancients and so-called primitive peoples can show us something astounding: that the universe is animate, and that far from being gross matter, the reality around us is "ensouled" – filled with volition and intelligence. The mythopoeic imagination – the human faculty employed in myth-making – joins us in a correspondence, a co-response of emotion and sympathy, with the universe around us. The recognition of this co-response is the foundation of the

The Stars, Oracles and Fate

The Stars, Oracles and Fate

In many cultures, flocks of birds have been accorded divinatory significance. This belief may, in part, be based on the observation of migrant birds, whose arrival heralds a change of weather.

universal phenomena of omen-reading and divination.

It is in the ancient civilization of Mesopotamia, the region of modern Iraq, that we find the most sophisticated development of divination. There is evidence from the 3rd millennium BCE of animal entrails being ritually examined in order to determine a god's answers to questions about the future. From at least 1900BCE liver

divination (or extispicy – examining the liver, lungs or colon spiral) was codified, with an extensive interpretive literature recorded on clay tablets. As well as divination through extispicy, the Mesopotamians employed lecanomancy (divination from oil patterns on water) and libanomancy (divination from the configurations of the smoke of burning incense).

Mesopotamian practices passed to Greece and eventually into the shadowy culture of the Etruscans of northern Italy, whose noble families held tenure of the College of Augurs in ancient Rome. The word augury has its root in the Latin *avis*, meaning "bird", indicating the status of bird omens in early times. The heavens were the home of the gods, and it appears that signs from the sky, such as birds and thunder and lightning, which were under the aegis of the supreme god Zeus or Jupiter, came to be elevated

An ancient Greek vase painting showing the famous oracle at Delphi. In the foreground, the tragic hero Orestes squats on the *omphalos*, or "navel", the sacred stone marking the world's centre.

The Stars, Oracles and Fate

The Stars, Oracles and Fate

An illustration based on a bronze model of a sheep's liver, used for divination by the Etruscans. The surface is divided into sectors of the heavens.

above signs observed on Earth. This helps to explain the importance of astral omens over all others in late Mesopotamian culture, certainly by the 8th century BCE. An increasingly accurate knowledge of the cycles of stars and planets, which represented the gods, gave the priestly possessors of this knowledge the semblance of a mastery of fate itself.

At its highest elevation, divination takes on the possibility of a transcendent self-knowledge. At Delphi in Greece, where the Pythian priestess sat at the navel of the world to speak the oracles of the Sun-god Apollo, there was a famous inscription: "Know thyself." Self-knowledge is the navel of fate, the axis around which our conscious being turns. This wisdom is the essence of divination, and is considered the highest knowledge of all oracles.

Divine Messengers

The reading of omens and portents from the sky gradually evolved into the divinatory art of astrology, with its attendant science of astronomy, in the first millennium BCE. However, long before then, the collating of observations of the sky, including meteorological phenomena and the appearances of the Sun, Moon and planets, had established the essential outlines of a symbolism that survived little changed for thousands of years. The Mesopotamian omen collection *Enuma Anu Enlil*, dating from c.1000BCE but incorporating much older material, gives brief formulaic readings on a range of sky phenomena, including eclipses, lightning and cloud formations.

Centuries later, omens from the sky were still a popular source of divination. "When beggars die there are no comets seen;/The heavens themselves blaze forth the death of princes." These lines from Shakespeare's *Julius Caesar* show

Divine Messengers

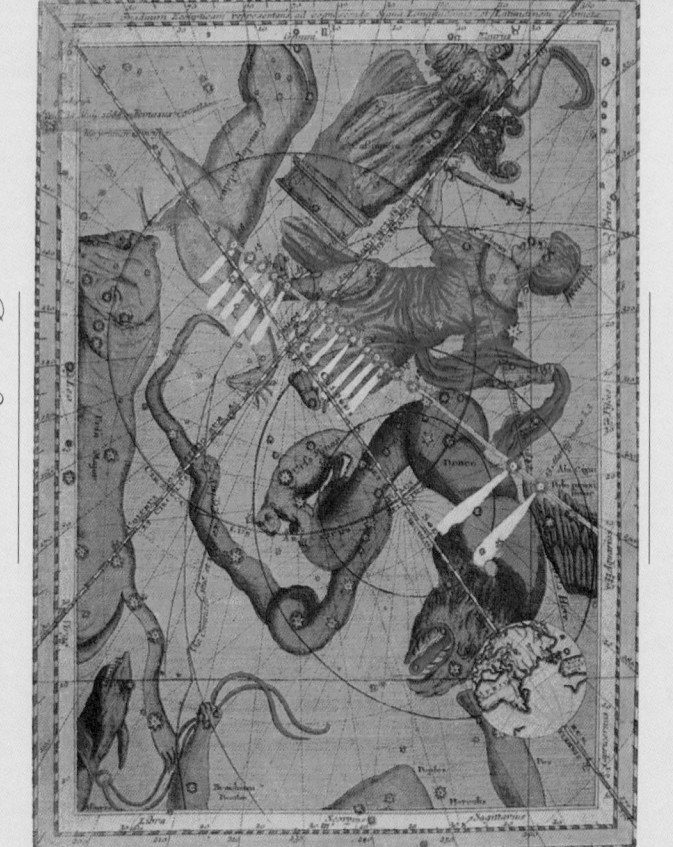

This illustration shows the comet of 1742 in various stages across the Northern Hemisphere sky. The comet travels across the constellations Camelopardalis and Cepheus to the wing of Pegasus. The Earth is shown at bottom right.

a remarkably consistent theme of interpretation from the earliest records to relatively recent times, linking dramatic celestial phenomena with social disorder, particularly the fall of rulers.

Among the most striking of all the heavenly portents are comets, which were universally interpreted as associated with the fates of leaders. From the standpoint of science, however, these are lesser members of the Solar System: huge snowballs of gas and dust moving on elongated orbits around the Sun, in cycles that can vary from a few years to many centuries. The brightest comets are an awesome sight, visible in the night sky for many months with a glowing head, a tail of gas and debris stretching away from the direction of the Sun for millions of miles. When cometary debris is scattered across the Earth's orbital path, it will eventually be captured by the Earth's gravitational

A drawing from 1618–19 that shows the movement of a comet over a period of 24 hours. The constellations across which it is travelling are Boötes and Ursa Major.

Divine Messengers

Divine Messengers

field and pulled into the upper atmosphere as meteors, or "shooting stars".

According to Ptolemy in the 2nd century CE, "[comets] show, through the parts of the zodiac in which their heads appear and through the directions in which the shapes of their tails point, the regions upon which the misfortunes impend."

Halley's Comet, with a period of 76 years, is the best-known comet in European history. Its appearance in 1066 was taken by William of Normandy as a favourable portent for his intended invasion of England – an omen recorded in the Bayeux tapestry.

One celestial portent that has never lost its potency, even in our secular era, is the Star of Bethlehem. Arabic and later European astrologers tended to accept the Christian interpretation that

A relief in the cathedral of Autun, France, shows the Magi dreaming of the advent of Christ. The angel indicates the star that heralds the birth of the Messiah.

the star was a supernatural sign from God, and therefore outside the realm of conventional astrology. However, in the early 17th century Johannes Kepler established the likelihood that the "star" was the conjunction of Jupiter and Saturn at the end of Pisces, near the equinoctial point, in 7BCE. Kepler felt that God might have also marked this conjunction with a nova, a new star flaring in the sky.

Recently, the English astronomer David Hughes has supported the hypothesis that Christ was probably born in 7BCE, and that the biblical reports are a layperson's version of what the conjunction would have meant for astrologers. Pisces is a sign of rulership for Jupiter, the planet of kings, and Saturn was the planet of the Jews. Hence, the conjunction in Pisces symbolically means "King of the Jews". Interpretation aside, it is an impressive illustration of the ancient understanding that the heavens themselves give signs of great events in the world below.

The Story of Astrology

From humankind's earliest speculations, what we now understand as the objective science of astronomy was inextricably bound up with astrology – the search for a transcendent meaning to our subjective experience, interpreting our destinies from the stars and planets.

Astrology developed from a complex amalgam of Babylonian and Persian astral religion and augury, Egyptian cosmology and calendar construction, and Greek scientific and philosophical speculation. Its classical form, including basic doctrines such as the order and interpretation of the twelve zodiac signs, were for the most part settled in the Hellenistic

This mosaic in the ancient Beth Alpha synagogue in Israel depicts the signs of the zodiac surrounding the Sun-god Helios.

period, when Greek civilization penetrated from the Mediterranean to northern India in the wake of the conquests of Alexander the Great after 334BCE.

Astrology, as we know it, depends on linking the positions of heavenly bodies, especially the planets, at a certain moment, with the circumstances of that moment. Like the science of calendar construction, divination by planetary movement requires accurate knowledge of astronomy, together with an established canon of interpretation of individual planetary qualities. In the 5th century BCE, these combined to produce the first Babylonian horoscopes – maps of the heavens for an exact time and place on the Earth. Judicial astrology – the practice of judging character and fate, especially from the horoscope cast for the moment of birth – developed from that time.

During its formative years, astrology seems to have become infused with a

The Story of Astrology

The Story of Astrology

tendency to fatalism. The heavens had always belonged to the great gods and, with the growing certainties of astronomical prediction, it was possible, by anticipating the movements of the planets, to predict the omens of the gods in advance. The Greek and Roman Stoic philosophers treated the universe as a whole, with every part in correspondence and sympathy with every other part; unsurprisingly, with only a few exceptions, the Stoics were significant allies of astrology.

Astrology achieved political influence in Rome, to the extent that in 27BCE,

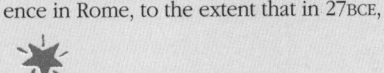

the emperor Augustus issued silver coins bearing the Capricorn glyph, believed to be his Moon sign. According to the historian Tacitus, his successor Tiberius used astrologers on several occasions.

However, astrology has not always been so well received. Its greatest early opponent was Christianity, because the practice of astrology maintained the survival of the old pagan gods. St Augustine of Hippo (354–430CE), who had studied astrology in his youth, demonstrated that it was founded on irrational arguments. Furthermore, he claimed that the astonishingly accurate answers sometimes given by astrologers were prompted by demons, who aimed to tempt the souls of both astrologer and client into a belief in stellar determinism – thus taking away the free will of the soul – and, worse, into the worship of planetary gods. His argument is sometimes used by Christians today.

Paganism was dying in the last days of the Roman Empire and astrology

A 14th-century Arabic depiction of the constellation Libra, from a copy of a manuscript by Al-Sufi. His interpretations of the constellations were based on the account by Ptolemy of Alexandria.

The Story of Astrology

could similarly have slipped away but for its capacity to disguise its religious origins and interweave its theory with the prevailing Greek-dominated science and philosophy. This was the achievement of Ptolemy of Alexandria (2nd century CE), who extended Aristotle's notion of change emanating from the rotation of the Celestial Sphere into a natural and easily comprehensible rationale of the subtle influence of the heavens – on the seed at the instant of conception and, by extension, at the moment of birth.

Ptolemy's rationalization eventually brought about a limited concord with Christianity at the time of St Thomas Aquinas (c.1225–c.1274). The Church declared that there need be no argument with astrologers provided that they limited that influence to the body, which might, in turn, influence the soul, but could not ultimately dictate its free will.

The Renaissance gave renewed impetus to astrology, which made its appearance at the highest levels of soci-

Astrologers (centre right) cast horoscopes during the celebrations of the birth of the Tartar leader Timur (1336–1405).

The Story of Astrology

ety and political influence. However, the rise of a scientific, objective imagination by the end of the 17th century, coupled with the overthrow of Ptolemy's geocentric cosmos by the Sun-centred system, made astrology an anachronism for the educated classes in Europe by the start of the 18th century. It has taken a new renaissance in the 20th century, and a scepticism about science as the sole answer to humanity's problems, to remind us that the core intuition of astrology, connecting us to an intelligent cosmos, may be just as relevant now as it was for the Mesopotamians, the Egyptians and the Greeks.

ALCHEMY

As in the heavens, so in the dense body of the Earth: alchemy brings the stars and planets into the metals, minerals and crystals where their powers may be released and purified by the initiate.

An esoteric and cosmological interpretation of the operations of early chemistry and metal-working – the starting-point for alchemy – appears to date from ancient Egypt. The Greek, Islamic and European alchemists attributed their magical knowledge to the mythical Hermes Trismegistos, the divine master of magic. The god and planet Mercury (identified with the Greek god Hermes) was said to be the living spirit of alchemy, with the power of transmutation.

From the 4th century BCE, alchemists established associations between the heavenly spheres and metals. Gold was associated with the Sun, silver with the Moon, quicksilver with Mercury, copper with Venus, iron with Mars, tin with

In this German woodcut of 1519, an alchemist looks to the sky to channel stellar energy into his furnace. At the side of the furnace, a powder is shown pouring from the funnel into a vessel – an indication that calcination is in progress.

Jupiter, and lead with Saturn. At any given time, the strength of a planet was said to affect the rate of growth of its associated metal – the interpretation of planetary movements was therefore vital to the alchemist's work. Furthermore, alchemists used astrology to determine when to begin new stages of work.

A vast range of metals and other substances was used in their work, which usually began by purifying the materials, turning them to powder through intense

Alchemy

Alchemy

heat (calcination). The goal, at the material level, was to transmute base metals, such as lead, into silver and, finally, gold. Many alchemists were also seeking the secret of organic life, the reward for which would be the elixir of immortality. This is an especially marked theme in the tradition of Chinese alchemy as practised by Daoist adepts.

To state simply that alchemists strived to produce gold loses sight of the spiritual dimension of their work. In early Greek writings, the accounts of material operations appear alongside a description of a magical or spiritual reality attuned to the soul of the alchemist. The tradition as a whole insisted that the art could be attained by grace alone. In alchemy, the stars and planets were a source of celestial energy that could be employed to "create" materially. By purity of motive only, the natural energy of the heavens could be channelled to turn base metals into gold and, spiritually, to discover the secret of eternal life.

SACRED CALENDARS

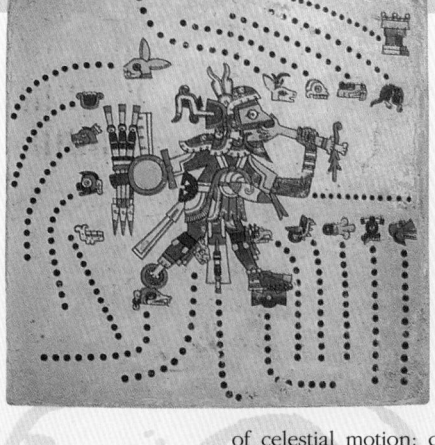

It is, perhaps, surprising to learn that up until the very recent past humanity understood, in a very practical sense, that time comes from the sky. It is none other than the record of celestial motion: day and night and the seasons are time.

An illustration from the Codex Féjérváry-Mayer of the chief god Tezcatlipoca eating part of a sacrificial victim. Surrounding him are the symbols of the twenty Aztec days.

In almost all cultures, the annual cycle of the Sun and the monthly phases of the Moon have been the foundation for the count of time beyond the day, but many different formulas have been used to arrange these basic elements. Every calendar has to cope with awkward fractions: for example, there are approximately 29.5 days in a lunar month and just under 365.25 days in a year. As these fractions accumulate, fixed festivals drift away from the phenomena they mark. The solution is intercalation, the periodic insertion of an extra day or other unit of time to realign a calendar with the heavens. Mesoamerican civilization developed a system of time-keeping on a number-base of 20, superimposed on an early solar calendar of 365 days. With the help of the calendar, the Maya achieved accurate astronomical computations. They established the synodical period of Venus – when the Earth, Sun and Venus come into the same alignment – as just under 584 days. Thus, five Venus synods (2,920 days)

A huge Aztec Sun Stone Calendar bearing the face of the Sun-god surrounded by the ages of humanity, then symbols of the days and stars.

coincided with eight solar years of 365 days; the Maya held a ceremony every eight years, when they fine-tuned their Venus calendar to allow for the time it had fallen behind the true Venus cycle.

Modern time-keeping is deeply influenced by the stellar worship of ancient Babylon and Chaldea in ways that are now hardly recognizable, although they are still reflected in many European languages (see table, page 94). The day of the Moon (Latin, *luna*) is immediately apparent; the Lord's day (Sunday) remains solar in German and English. French, Spanish and Italian show us the planets Mars, Mercury, Jupiter (or Jove, Greek Zeus) and Venus for Tuesday, Wednesday, Thursday and Friday. English gives Saturday to Saturn, while English and German use Teutonic equivalents of the Roman gods for Tuesday (Tiwaz), Thursday (Donar), and Friday (Frea). The English Wednesday is from Wodan (Anglo-Saxon Woden), who was identified with Mercury.

Sacred Calendars

Sacred Calendars

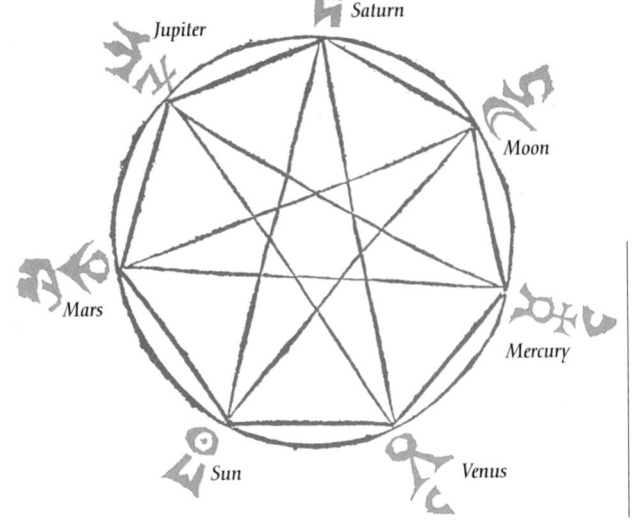

Planetary gods rule weekdays in an occult system predating 100CE. The planets are set in a circle in "Chaldean Order", from Saturn, slowest and furthest from Earth, to the Moon, swiftest and closest. Beginning at the Sun, draw a line to the Moon, then from the Moon to Mars, and so on until you complete the week and the 7-pointed star.

DAYS OF THE WEEK

Latin	French	Italian	German	English	Planet
Solis dies*	Dimanche	Domenica	Sonntag	Sunday	SUN
Lunae dies	Lundi	Lunedì	Montag	Monday	MOON
Martis dies	Mardi	Martedì	Dienstag	Tuesday	MARS
Mercurii dies	Mercredi	Mercoledì	Mittwoch	Wednesday	MERCURY
Iovis dies	Jeudi	Giovedì	Donnerstag	Thursday	JUPITER
Veneris dies	Vendredi	Venerdì	Freitag	Friday	VENUS
Saturni dies	Samedi	Sabato	Samstag	Saturday	SATURN

*later *dies Dominica,* "Lord's day"

CHINESE ASTROLOGY

The fact that traditional Chinese celestial sciences have parallels with Greek, Persian and Indian sciences suggests not only that they may have common archaic origins, but also that there may have been some cross-fertilization in historical times. However, these shared characterisitcs are fused with distinctively Chinese developments.

The Western and Indian traditions are rooted in the ancient Mesopotamian and Egyptian observation of the horizon, through the heliacal (with-the-Sun) rising and setting of stars and planets, and the ecliptic, the Sun's annual path through the constellations. Chinese astronomy and astrology, on the other hand, are both circumpolar and equatorial: they observe the northern circumpolar stars (which never set at the latitudes of China) and, in particular, the passage of the stars across the meridian (the great circle through the Pole and the observer's zenith, the point directly overhead; see pages 22–23).

A system of 28 lunar mansions, or *xiu* – in effect segments of the sky radiating out from the Pole – was established along the equator in the 1st millennium BCE. This approach, based on equatorial Moon stations, is found in Babylonian astronomy before 1000BCE, suggesting an early common origin. The *xiu* marker stars were selected according to their approximation in RA (Right Ascension, the

An image from a Chinese geomantic almanac. Geomancy (Earth omens) is interwoven with astrology in the Chinese system.

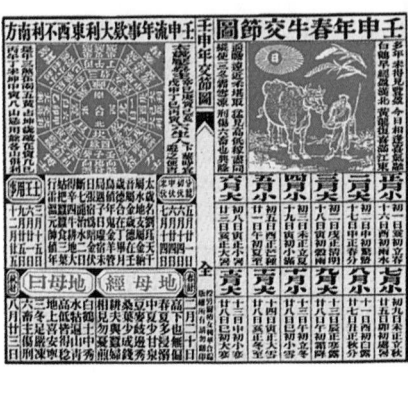

THE FIVE AGENTS AND THEIR COSMOLOGICAL ASSOCIATIONS

	Wood	Fire	Earth	Metal	Water
palace	spring	summer	late summer	autumn	winter
planet	Jupiter	Mars	Saturn	Venus	Mercury
climate	wind	heat	humidity	dryness	cold
number	8	7	5	9	6
colour	green	red	yellow	white	black
taste	sour	bitter	sweet	pungent	salt
smell	rank	scorched	fragrant	rotten	putrid
viscera	liver	heart	spleen	lungs	kidneys
animal	fowl	sheep	ox	horse	pig
emotion	anger	joy	sympathy	grief	fear

Chinese Astrology

measure along the celestial equator) to the dividing lines radiating out from the Pole. Circumpolar stars correlated with the equatorial Moon stations; these stars were the abodes of the heavenly bureaucrats who oversaw the earthly administration of the emperor, the "Son of Heaven".

The 28 *xiu* are divided into four equatorial "palaces" of seven *xiu* each: the Green (or Blue) Dragon (east and spring), the Vermilion Bird (south and summer), the White Tiger (west and autumn) and the Black Tortoise (north and winter). The circumpolar sky forms the fifth "Central Palace". This division brings astronomy into harmony with the ancient symbolism of the "Five Elements (or Agents)", found everywhere in Chinese occultism (see table, opposite).

Horoscopes giving planetary positions as a means of reading character or fate are not characteristic of Chinese astrology until relatively late, reflecting an importation from the mature traditions of Indian and Islamic astrology.

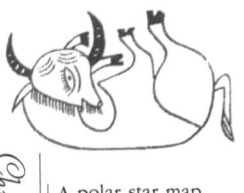

Chinese Astrology

A polar star map, *c*.940CE. The Plough, or Big Dipper, can easily be identified.

The popular version of the "Chinese zodiac" takes one element of Chinese astrology and applies it to years, symbolized through a cycle of animals: rat, buffalo, tiger, hare, dragon, snake, horse, goat, monkey, cock, dog and pig. The origins of the Chinese animal zodiac remain a mystery, and despite attempts to derive it from the zodiac of Western tradition, a common link has not yet been found.

MYTH AND THE SKIES

The projection of mythical figures onto the skies has a complex foundation, posing a problem of interpretation for modern scholarship. The perplexity emerges as soon as we look beyond practical elements, such as the needs of hunters or farmers to keep track of the Moon and the seasons.

The mythical accretions imposed on these practicalities represent a complex mixture of human aspirations, intimately bound up with religion, myth and poetry. It seems easier to deal objectively with Egyptian calendar-construction than it is to be sure what the experience meant for the Egyptian who depicted the Sun as Ra, seated in his boat on the back of the sky-goddess Nut.

In the 18th-century European Enlightenment, the worldview of the ancient Egyptians was perceived as being of purely antiquarian interest. The

Myth and the Skies

belief of an earlier age that myths were allegories of nature or the soul was shaken off. In place of interpretation, myths were classed as bizarre amplifications of spurious causal connections arising from poor observation or faulty data. Individual myths were sometimes explained as elaborations of historical events – an approach known as euhemerism, after Euhemerus of Messene (*c.*300BCE), who taught that the Greek gods were elevated kings and heroes.

Interpretations of myth have taken a variety of routes beyond this rationalistic impasse. The Scottish scholar James Frazer (1854–1941) exemplifies the view that myths, especially fertility motifs, are survivals of an earlier phase of cultural evolution which reflect humanity's struggle to master nature, first through magic, then through the gods and reli-

The Green Lion devouring the Sun, an alchemical symbol. The great psychologist Carl Jung saw the symbols of alchemy as part of the symbolic quest toward the whole self.

gion. The "historicist" interpretation, on the other hand, relates myth to real processes of history – such as the replacement of one theological system by another through invasion or cultural assimilation. The astronomical interpretation of myth (see pages 66–67) parallels the view that myths reflect humanity's response to nature.

Distinct from both the naturalistic and historicist trends is the 19th-century Romantic interpretation, which sees myth as an imaginative vehicle whereby humankind gives symbolic expression to spiritual ideals. This kind of "mythopoeic" creativity is not confined to civilizations like Egypt and Greece. The French anthropologist Lucien Lévy-Bruhl (1857–1939) suggested that the "primitive" mind is capable of understanding ordinary logical categories such as causation in our terms, but has not lost the capacity, as we in the modern world have, to perceive directly and respond to a supernatural realm. Omens, gods and spirits are real

Myth and the Skies

and their world is the world recorded in myth. The Mythopoeicism was overshadowed by the structuralist approach of Claude Lévi-Strauss (born 1908), who argued that myths reflect underlying patterns of social and cultural organization. However, it was Lévy-Bruhl's concept of mythopeic thought, coupled with the Romantic attitude to myth as spiritual revelation, that laid the foundations for the work of Carl Jung (1875–1961).

As illustrated by his study on precession (see pages 36–37), Jung's work develops the astronomical interpretation by relating it to a subjective dimension – the workings of the unconscious mind. He interprets the symbolism assigned to stars and planets as a residue of the "collective unconscious", a part of the mind that responds to certain in-built universal symbols, or archetypes. Symbol systems such as astrology and alchemy enable us to create a language of the soul's journey toward its mystical fulfilment or "individuation" – the process of forging the self.

Jung believed that the twelve character types depicted in the zodiac signs, together with the dynamic attributes of the traditional planets bringing movement and change into the picture, made astrology "the psychology of antiquity".

A detail from a 9th-century manuscript showing Christ in the guise of the Sun surrounded by the signs of the zodiac.

Myth and the Skies

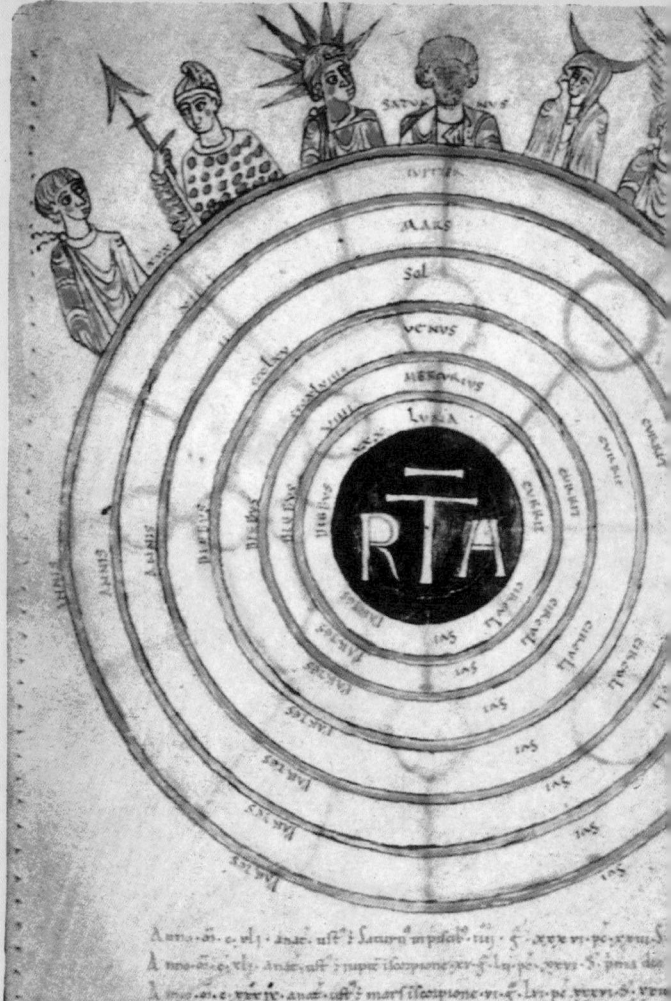

An illustration from the Dijon Bible, showing the planets and their spheres according to the Ptolemaic view of the cosmos: in Ptolemy's system, the Earth, not the Sun, was at the centre. The order shown here is, radiating outward: the Moon, Mercury, Venus, the Sun, Mars, Jupiter and Saturn. Personifications of the planets sit across the top, with Saturn flanked by the Sun and Moon.

SUN, MOON AND PLANETS

This chapter shows how the Sun, Moon and planets have exerted a hold over the imagination since ancient times. The Greeks knew only the five inner planets, which, with the Sun and Moon, were called "wanderers" (Greek *planetes*). European and Indian astrology have sustained a rich planetary symbolism into modern times, and we even see traces of the mythological process at work in the "new" planets, Uranus, Neptune and Pluto.

The Sun held aloft in the heavens, from a 14th-century manuscript. Across all cultures, solar symbolism suggests life, fertility and creation.

THE SUN

Nowhere else in the heavens is the contrast between science and myth more marked than in our conceptions of our greatest heavenly body, the Sun. This one phenomenon has produced two distinct truths through time.

In scientific understanding the Sun is an unexceptional star, just one out of thousands of millions of stars tenanting the inconceivable vastness of our patch of the universe. This incandescent sphere of mainly hydrogen gas, in permanent nuclear fusion at its core at temperatures in the region of 20,000,000°C (73,000,000°F), is source and sustainer to a system of planets orbiting around it. The planets, together with asteroids and various comets, are the coalesced remnants of material spun off at the time of the Sun's own formation from an enormous supernova explosion in our galaxy, around 4,000 million years ago. Everything else in the Solar System is a

A spiral Sun-symbol from an engraved stone found in Gotland, Sweden (5th century CE). Motifs such as these were common in northern Europe during the Bronze and Iron Ages.

fragment by comparison – the Sun is 1.3 million times the volume of the Earth, and a third of a million times its mass.

The dependence of Earth-life on the Sun is total. Science becomes infused with poetry when we consider the perfect balance of life on Earth, delicately attuned to the light and heat of the Sun.

Although the Sun must have impressed itself on human imagination from the dawn of consciousness, there appears to be a distinct evolution in its symbolic or mythic treatment. Symbolic suns have occasionally been found at prehistoric sites such as the Matopo Hills in Zimbabwe, but this is not generally a striking feature. Most cave paintings of Upper Paleolithic times (*c.*40,000 years ago) do not offer any representations of

the Sun: their art is focused on female fertility and animals of the hunt.

A distinct evolution in the symbolic meaning of the Sun seems to arise with an advance in settled civilization. This change can be traced in the myths of several cultures in which the Sun begins simply as one minor mythic character among many. For example, in Greek myth, Helios, the Sun, the offspring of the Titans Hyperion and Theia, was given inferior status. We know that he was thought to travel westward across the sky, heralded by Eos, the dawn, but there are few myths about Helios himself. Only later did great Apollo subsume the attributes of Helios to become the main Olympian Sun-god.

One of the earliest civilizations to establish an enhanced status for the Sun was ancient Egypt, from at least the third millennium BCE. Ra, the Sun-god, was the supreme creator deity, who took different forms when he rose (Khepri) and set (Atum). He was also Horus, the fal-

The Sun

con-headed god later identified by the Greeks with Apollo.

Khepri, the Sun-god at dawn, was represented as a scarab beetle, who rolled the ball of the Sun over the horizon. The hieroglyph for the scarab eventually evolved into the astrological glyph for Cancer, the sign associated with the midsummer solstice. The glyph symbolizes for later times, as it did for the ancient Egyptians, the perpetual fertility and renewal of life.

The culmination of the Sun's status in ancient Egypt was reached in the brief religious revolution of Pharaoh Akhenaten ("Glory of Aten") in the 14th century BCE, who exalted the Sun-disk (Aten) above all other gods. This idea was to recur in later centuries, most impressively with the solar cults that developed around the Roman emperor.

The story of Sun-worship in Rome interweaves different strands of celestial symbolism, often serving the purposes

A Babylonian terracotta statuette (*c*.2000–1750BCE) thought to represent the Sun-god Shamash, the son of the Moon-god Sin and brother of the love-goddess Ishtar. The three represent the divine Babylonian triad.

of political propaganda. The first strand was the cult of Mithras, imported into Rome from Persia. Mithras was a bull-god, linked with the constellation Taurus (see pages 225–228); he was often shown banqueting with the Sun. Yet in a paradoxical inversion so characteristic of mythology, he was also depicted as a solar god who slew the bull. In this form he was known as Helius, the Sun-god, and Sol Invictus, the Invincible Sun.

The second strand derived from the cult of the Phoenician Sun-god Baal, who was worshipped in the form of a black stone. Baal became popular in the Roman Empire in the 2nd century CE. In 218CE, when Elagabalus became emperor as Sol Invictus Elagabalus, the cult of the Sun was founded as an official religion. Aurelian (reigned 270–275CE) adapted the solar cult to accord better with traditional Roman religion under the title Deus Sol Invictus – God, the Invincible Sun. This lasted until the reign of Constantine (reigned 312–337 CE),

The Sun

when Christianity took root, banishing (and at the same time assimilating) its solar rival. The festival of Sol Invictus was later celebrated on December 25, the date adopted by the Christians for their own invincible king.

Sun-worship reached its most dramatic form in central American civilization. The Aztec epic of creation ends with the generation of the Fifth Sun, following the four previous eons, the Suns of earth, wind, fire and water. In this culture the gods themselves must be sacrificed to make the Sun move. One by one, the feathered-serpent god Quetzalcoatl cuts out their living hearts with a knife, and by this act the moving Sun, Nahui Ollin, is created. This is the basis for the terrifying human sacrifices to the Sun carried out by the Aztecs.

The Sun symbolizes truth and integrity: "know thyself" runs the motto at the oracle sacred to the god Apollo at Delphi in Greece. But the apparent brightness of the Sun conceals a

Nefertiti, the queen of Akhenaten, making an offering to the Sun-god, shown as the solar disk Aten from which hangs the *ankh*, symbol of the rising Sun.

mystery. The great neo-Platonic philosopher of the Italian Renaissance, Marsilio Ficino (1433–1499), taught that we "see" with two faculties, one of the concrete mind of ordinary thought, and one of the higher intellect. In his poem *De Sole* ("On the Sun"), his last major work, Ficino shows that the Sun has not one but two lights: the ordinary light of earthly senses, and a hidden, occult light – the inspiration of astrology.

This idea of the "hidden light" of the Sun is found among the Pueblo peoples of North America who teach that Oshatsh, the physical Sun, despite his blinding brilliance, is a shield to protect humankind from the light of the one Great Spirit. Such profound intuitions, which are consistent with our own Western preoccupations, remind us of the relevance and subtlety of thought found in innumerable myths of many non-Western cultures.

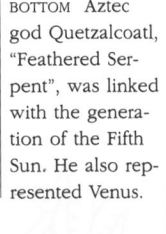

BOTTOM Aztec god Quetzalcoatl, "Feathered Serpent", was linked with the generation of the Fifth Sun. He also represented Venus.

The Sun

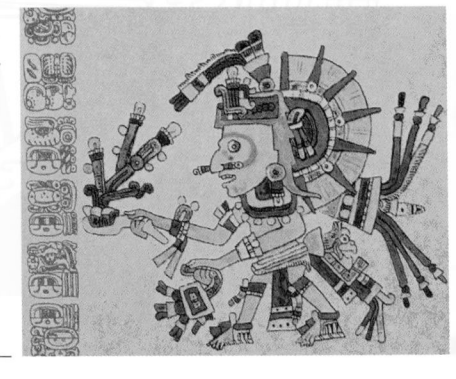

THE MOON

Having a diameter of 2,160 miles (3,476km) compared with the Earth's 7,927 miles (12,714km), the Moon is the largest satellite in the Solar System. Earth and Moon are gravitationally yoked to each other so that the Moon's rotational period around its axis exactly matches its orbital period around the Earth (27.32 days): this means that we always see the same face of the Moon and that its far side is hidden from our gaze. However, our satellite is a dead world, with virtually no atmosphere, no surface water, and no chance of life as we know it.

As with the planets, we are able to see the Moon only by reflected light. The crescent of the New or Old Moon, or the complete disk of the Full Moon, is illuminated directly by the Sun, giving a cycle of phases lasting 29.53 days, waxing from New through first quarter to Full, then waning through third quarter to its disappearance at the "Old Moon", ready for the next New Moon. A beautiful phenomenon in the evening sky known as "earthlight", or the "Old Moon in the New Moon's arms"; is the shadowy effect produced when sunlight reflected from the Earth falls on the dark part of the lunar disk.

The symbolism associated with the Moon across different cultures and times is often complex and paradoxical. The Moon appears to have had greater eminence than the Sun in prehistoric times, and it is believed that in most cultures the calendar started as a count of lunar months rather than solar seasons. In a similar vein, many astronomical megalithic sites were erected to track the Moon's orbit.

From our perspective, the Sun and Moon are a heavenly duo, the two great lights of the sky, almost universally seen as the twin rulers of day and night. There is also the extraordinary circum-

A Tarot card representing the Moon. In Tarot divination, the Moon is said to foretell hidden perils, illusion and deception.

The Moon

stance that, despite their vastly different distances, they both appear to be the same size.

The ancient Egyptian Moon-god Thoth shows an ancient priestly interpretation of the fact that as the Sun and Moon rise and set, they replace one another in the sky: as the Sun-god Ra made his underworld journey in the hours of darkness, Thoth was required to take his place in the upper world. Thoth was also responsible for regulating the calendar, taught humanity the arts and sciences, and was interpreted by the Greeks as the god Hermes. In a later epoch, the Moon-god Thoth (as Hermes Trismegistos) became the inspiration for the "hermetic" tradition of Greek, Islamic and European occultism.

The Moon's function as regulator over the menstrual cycle (from Latin *mensis*, "month") gave it an association with fer-

The Roman goddess Luna, from *Mythologiae* by Natalis Comitis (1616), shown with her attributes: the crescent Moon on her head, the torch in her right hand, a bow and arrows on her back, and a dog at her feet.

tility in ancient times, and the Moon appears to have been allotted an increasingly feminine role, with the Sun taking the masculine part. A characteristic illustration of the feminine Moon in her most benign aspect is provided by the Moon-goddess Chang E, or Heng E, one of the most popular figures in Chinese folk belief. The Festival of the Moon, held on the Full Moon following the autumn equinox, is one of the three great annual feasts.

Today we perhaps tend to take for granted the feminine symbolism of the Moon, yet it has also notably been treated as masculine – as in its embodiment as Thoth in ancient Egypt. The Japanese lunar god Tsuki-Yomi is male and in early Mesopotamian mythology the Moon-god Sin was an old man with a beard.

In Hindu mythology, the Moon is said to be where departed souls go. The notion of the Moon as a realm of the dead brings us to a further tension in its symbolism. Its phases can suggest an

The Moon

The Moon

analogy with organic cycles of birth, growth, decay and death. In the mythology of parts of South America, the Moon is thought to be the mother of grasses. In ancient Mesopotamia, some considered the heat of the Moon, rather than the Sun, to be the energy force by which plants grew. This life-death paradox is contained in the Moon as Triple Goddess, a myth-motif that appears in many guises, especially where we find a female trinity, as in the three Fates, or the three witches. Ancient Greek poets saw the virgin huntress Artemis (the Roman Diana) as a goddess with three forms, the other two aspects being Selene, the Moon in the sky, and Hecate, a mysterious goddess of the lower world. The Triple Goddess can be interpreted as three phases of the lunar cycle: the silver bow carried by Artemis represents the crescent New Moon, Selene is the mature Full Moon, and Hecate is the dark of the Moon – some-

The Moon, like the Sun, has been depicted as a figure borne by a chariot across the sky, as in this roundel from a 13th-century stained-glass window. Luna is Latin for "Moon" (see also p.116).

A painting on a Greek bowl of *c.*490BCE showing the Greek Moon-goddess Selene in her chariot. The daughter of the Titans Hyperion and Thea, Selene became the lover of Endymion, the king of Elis. She wanted him to live for ever, and put him into a magic sleep in which he remained eternally young.

times she was depicted as a crone, showing the last moments of the Moon's cycle, and libations were offered to her at the end of every month.

The phases of the Moon are accounted for in mythology around the world. In Maori myth the (male) Moon kidnapped the wife of the god Rona. Enraged, Rona confronted the Moon, and now they are forever locked in battle in the sky. As the Moon wanes it is said to be growing weary of fighting and needing to rest, which it does during the waxing period of the cycle; when it is full, the fighting commences once more.

In psychological and astrological symbolism, the Moon stands for the subliminal realm – the subdued light of the unconscious, in contrast to the brighter illumination of the conscious mind.

The Moon

MERCURY

Mercury

Mercury, which is less than half the diameter of the Earth, is the smallest of all the planets. It is also the closest to the Sun, at 36 million miles (58 million km). Viewed from the Earth, Mercury never appears to be at a distance farther than 28° from the Sun, and over its 88-day orbit it seems to shuttle back and forth (through retrograde motion; see pages 34–35) in attendance upon the great luminary, as befits its ancient mythological interpretation as herald or messenger. However, its closeness to the Sun obscures it from casual naked-eye observation, and it can only be satisfactorily seen, briefly, at sunrise or sunset in early spring or autumn.

For the astrologer-priests of ancient Mesopotamia, Mercury was the god Nabu, and in honour of him a cult was established with its principal centre at Borsippa, a city to the south of Babylon. There are now only sparse accounts of

A 15th-century depiction of Mercury. He holds a bag of coins, indicating his role in commerce, and the caduceus, a magical wand entwined with two snakes that was said to have healing powers. In the small circles are the zodiacal signs Gemini (left) and Virgo (right) which Mercury is said to influence.

Mercury

this cult, but it is known that *c*.1000BCE Nabu replaced an early Sumerian goddess Nisaba (or Nidaba) as the patron of scribes. Nabu was the son of royal Marduk (equivalent to the Roman Jupiter),

Mercury

and any variation in the condition of the planet Mercury presaged change for the son of the king, the crown prince. However, Nabu's most important function was as the scribe of the gods. On the seventh day of the spring festival marking the Mesopotamian New Year, he rescued Marduk from captivity, symbolizing a restoration of authority and order for the year to come; and on the eleventh day the gods gathered to decide the destiny of the world, while Nabu recorded their judgments.

The Roman god Mercury was originally a god of trade, and words such as "merchant" and "commerce" derive from the Latin root of his name. He came to be identified with Hermes, the Greek god of motion, transfer and exchange. Hermes was originally a phallic fertility god and the god of travellers. His name literally means "he of the stone heap": the god was honoured by wayside cairns to which travellers would add stones. Hermes guided the souls of the

A drawing by Henrik Goltzius (1587) of Mercury as a youth in a winged cap (an alternative to his winged sandals). According to Julius Caesar, a god he called "Mercury" was the most widely worshipped god of the Celts in Gaul and Britain.

Mercury

dead to the underworld and was a messenger of the gods.

The son of Zeus and the nymph Maia, Hermes' divine nature was established on the day of his birth. Before noon on that day he had invented the lyre, and before the day was out he had stolen the cattle of Apollo, his half-brother. Apollo was furious, but Zeus was charmed by the clever child and made him his cupbearer. Some of the oldest depictions of Hermes show him as an old man with a long beard, but usually he was depicted as a beautiful youth.

In the late centuries BCE, Hermes came to be equated with the Egyptian god Thoth. Prior to this, Thoth had been a Moon-god, albeit with many of the characteristics attributed to Hermes. He was the patron of learning and also the messenger and scribe of the gods. This shift in attribution from the Moon to Mercury is an example of cultural assimilation under growing Greek influence from the 4th century BCE.

Mercury

This same period saw the distinctive naming of Hermes-Thoth as Hermes Trismegistos ("Thrice-Greatest Hermes"), an aspect of the god's symbolism in later magic and alchemy. This god was credited with passing on to humankind medicine, magic, astrology and alchemy. In European alchemy, we find him in his Latin form, Mercurius – the goal of alchemical work and the secret guide to the adept, sometimes a Christ-figure and sometimes a trickster or a dragon guarding the secret of the Philosopher's Stone.

In astrology, it is believed that those born under the influence of the planet Mercury are quick-witted, crafty, alert, and able to think and talk lucidly and fluently. However, they are also believed to have a tendency to fickleness.

VENUS

Venus is sometimes referred to as a twin
sister of the Earth, and there is a kinship
in the purely physical dimensions of
both planets. Venus is close in diameter
to the Earth, at 7,600 miles (12,231km)
compared with the Earth's 7,927 miles
(12,757km), although it is lighter, having
seventeen per cent less mass. Venus
revolves around the Sun in 225 days at a
distance of 67 million miles (108 million
km) inside the Earth's orbit. The effect
of this from the Earth is that Venus, like
Mercury, appears to travel through the
sky close to the Sun, its maximum elongation never exceeding 48°. One day on
Venus is equivalent to 243 days on
Earth, because Venus takes longer to
revolve on its own axis than it does to
orbit the Sun. Venus revolves in the
opposite direction (east to west) from that
of most of the other planets in the Solar
System, setting itself apart not only by
virtue of its exceptional beauty, but also

Venus

owing to its physical characteristics.

A planet's visibility depends on its ability to reflect light from the Sun. The dense cloud atmosphere surrounding Venus permits a large proportion of the light falling on the planet

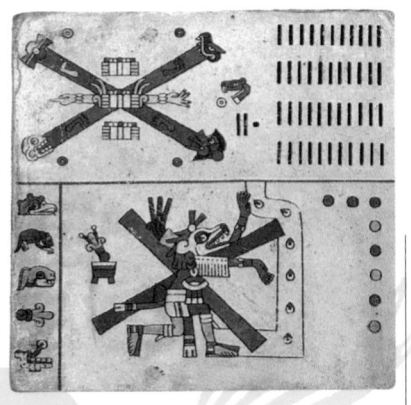

to be reflected and, as a result, at certain phases Venus will outshine any object in the sky apart from the Sun and Moon. On a moonless night with good weather, it is possible to observe the shadows cast by the planet's ethereal blue-white light.

Cross-cultural comparisons of the symbolism of Venus reveal as many differences as similarities. The light of Venus, which to most Europeans seems so beautiful, is viewed with caution in traditional Chinese astrology. Venus was called "Grand White" on account of its

Venus is alternately seen as the Morning or the Evening Star. This illustration from the Codex Féjérváry-Mayer shows the Aztec god Xolotl, Venus as the Evening Star, at the crossroads of fate. The god later became the mythological twin of the supreme god Quetzalcoatl, the Morning Star.

colour, but in China white is often considered unlucky, ominous and ghostly. Wherever Grand White appeared in the sky, it was taken to signify arms and punishment. This connection stems from its association with the negative, dark, *yin* principle, as distinct from the positive, light, *yang*. The appearance of Grand White in daylight (which may happen just after sunrise or sometime before dusk when Venus is near to maximum elongation) implied the *yin* overcoming the *yang*, and therefore trouble for the sovereign from the lower orders.

The warlike aspect of Venus is particularly evident in Mayan culture. Decisions on when and where to do battle were timed to the cycles of Venus and

Venus

An ancient Roman stone relief depicting the birth of the goddess Venus (equivalent to the Greek goddess Aphrodite). Venus was born, fully formed, from the foam where the castrated genitals of the god Uranus fell into the sea.

Venus

Jupiter. However, this is only one facet of the leading role played by Venus in Mesoamerican culture. The most important of all the gods, the plumed serpent Quetzalcoatl, was identified with Venus as Morning Star, rising in the east just before dawn. The Mayan ritual calendar system was integrally linked with observation of the planet. Mesoamerican skywatchers had at an early period gained an exact understanding of the 584-day synodic cycle of Venus as the planet goes through its phases of Morning and Evening Star. The Venus table records 65 of these 584-day cycles: in all, 37,960 days. This coincides with 146 of the 260-day counts of the Mayan calendar, and 104 Mayan solar years of 365 days.

These exact multiples within the Venus cycle elevated the planet's standing among the Maya above that of any other. At his heliacal rising (when he is first observed rising just before the Sun), the planet-god Quetzalcoatl is shown throwing the spears of his dazzling rays to impale his enemies. In some parts of ancient Mexico, people would lock their windows and doors to protect themselves from the rays of Venus, which they believed would bring disease and death as it rose alongside the Sun.

The Mesopotamian interpretation of the heavens gives us glimpses of a very different Venus from the planet that, in Western culture, is associated with love. The original Sumerian deity linked with Venus was Inanna, also known as Ishtar, part of the great Mesopotamian triad of the Sun, Moon and Venus. As well as the goddess of love, sex and fertility, Ishtar was also a fierce war-goddess known as the "Lady of Battles". Ishtar's Greek equivalent, Aphrodite (the Roman Venus), was the goddess of love and sexuality, whose most famous lover was the war-god Ares (Mars).

A medieval woodcut depiction of the planet Venus. The figure is seen with the zodiacal constellations Libra (left) and Taurus (right), over which she rules. The planet's glyph is shown on the right of the picture, beneath the emblem of the Morning Star.

Venus

MARS

Mars

At 142 million miles (229 million km) from the Sun, and with an orbit of 688 days, Mars is the first planet beyond our own. It is relatively small, a little over half the diameter of the Earth, but its closeness to us ensures that it appears quite bright. It has two moons, Deimos and Phobos, the latter being so close to Mars that it orbits the planet in 7.5 hours. The atmosphere of Mars is similar to that of the Earth. It seems to have water and oxygen and its surface temperatures can reach 77°F (25°C) – it is therefore possible that it sustains some form of life.

For several centuries, Martian features seen through a telescope have excited hopes of an extraterrestrial civilization. The most important of these features are the "canals", lines which appear too regular to be natural. But recent satellite probes

have established that, apart from a huge canyon, the lines are an optical illusion. More recent speculation has surrounded pyramid-like formations that are clearly visible from satellite pictures.

Cross culturally, there is more consistency in the symbolism of Mars than in the other planets, perhaps owing to the distinctive red hue produced by the iron oxide on its surface: red is universally associated with fire and blood.

In Chinese tradition, the red planet is associated with the element fire, the heat of summer and the heart. The Daoist alchemists saw it as cinnabar, red mercuric sulphite. From the time of the first civilizations of Mesopotamia, Mars has taken on a malevolent identity as lord of the dead, bringer of pestilence and war. His earliest known form was the Sumerian Lugalmeslam, ruling the lower world to which the Sun travels every night. Early on, this deity seems to have become assimilated to Nergal, the main representative of Mars for the

OPPOSITE The planet Mars from an English "shepherd's calendar" of 1497. He is seen with fire and the zodiacal signs Aries and Scorpio, over which he rules.

Mars

Mars

Mesopotamians. Called to account for insulting the emissary of Ereshkigal, the goddess of the underworld, Nergal was ordered to descend into her realm; but he was forewarned by Ea, the god of wisdom, to accept nothing from her.

Ereshkigal's attendants brought a chair, but he would not sit; they brought him bread and beer, but he would not eat or drink; they brought him a foot-bath, but he would not wash. Ereshkigal then went to bathe, and allowed Nergal a glimpse of her body; at first he resisted, but when the goddess repeated the performance, Nergal gave in.

They made love for six days and on the seventh day Nergal went back to the upper world. Ereshkigal threatened to raise the dead until there were more dead abroad than living unless her lover was returned. The not-unwilling god was then pulled back to share rulership of the underworld with Ereshkigal.

There is a related Mesopotamian tale of Ishtar descending into the under-

A fresco found in Pompeii and dating from the 1st century CE, showing the war-god Mars in his battle robes embracing Venus, the goddess of love.

world, and the pairing of Nergal and Ishtar recalls the later symbolism of Mars and Venus. In Classical myth, Nergal was readily assimilated to the Greek war-god Ares, who for the Romans was Mars. The Greeks held Ares in low account, but the Romans exalted Mars as a military god. Ares/Mars loved war for its own sake, and was indifferent to the rights and wrongs of the parties for whom he fought.

Mars

Most of the other gods despised him, except for Eris, the goddess of strife, and Aphrodite (Venus). The relationship between Mars and Venus was unashamedly erotic, and their pairing has come to personify sexuality and sexual

Mars

difference. Even in modern science, the glyphs for the two planets are those for male and female (see page 8).

Mars is at the heart of the most significant modern scientific astrological research undertaken, by the French statistician and psychologist Michel Gauquelin (1928–91). His first breakthrough concerned the "Mars effect" in the early 1950s. In carefully controlled experiments, he drew data from thousands of French birth certificates, on which both time and place of birth are recorded. From this material, he established, with considerable statistical evidence, that French solo athletes tended to be born with the planet Mars in four significant horoscope positions.

The presence of Mars in these birth charts reflects the strong will and single-mindedness that are required to achieve at a high level in competitive sport. Hostile scientists unsurprisingly dismissed Gauquelin's "Mars effect" as a fluke, but his method has never been faulted.

JUPITER

A medieval drawing of Jupiter. The fish and arrow indicate Pisces and Sagittarius, the signs he governs.

The giant of the planets, Jupiter is eleven times the diameter of the Earth and more than three hundred times its mass, with tremendous gravitational and rotational pressures creating in its dense atmosphere a maelstrom of methane and ammonia. With an orbital radius of 483 million miles (777 million km), Jupiter is scarcely touched by the distant Sun's light and heat, and life as we know it is unimaginable. This planet's identity as king of the gods befits its stature in the Solar System, with its retinue of twelve moons; Galileo discovered four of these in 1610, and they may be seen with good binoculars.

Jupiter's period of revolution around the Sun is 11.86 years. The approximation of this cycle to twelve years means that the planet marks out one sign of the zodiac every year, just as the Sun marks out one sign every month. This

Jupiter

Jupiter

correspondence of motions led early Chinese astrologers to term Jupiter the "Year Star". The planet was understood to generate power in each star-group through which it passed and, as in European astrology, it was deemed a divine law-maker, attuned to the human administration of authority in the world below. In traditional Chinese illustrations, Jupiter appears as the magistrate, the local representative of imperial authority.

In Mesopotamian myth, Jupiter was the planet of Marduk, the patron god of Babylon. Marduk appears to have started life as an agricultural deity who was associated with the fertilizing power of water. This idea can be traced forward to the Romans, who honoured their god Jupiter as Jupiter Pluvius, the bringer of rain. In the Mesopotamian epic of creation (2nd millennium BCE), the first gods were

This boundary stone (12th century BCE), shows Mesopotamian gods including Ishtar (top left). In the centre are the dragons of Marduk and Marduk's son Nabu.

Jupiter

created from the primeval pair, the god Apsu and the goddess Tiamat, but their offspring were so noisy that Apsu wanted to destroy them. Ea, one of the brood, learned of this and slew Apsu. Tiamat, stirred to vengeance, marshalled a troop of hideous monsters. Marduk, the son of Ea, agreed to save the gods from Tiamat if they promised him supreme authority.

The gods set up a constellation in their midst and asked Marduk to destroy and recreate it at will. He did so, and the gods rejoiced and called him king. He captured Tiamat's monsters and then murdered the goddess. Marduk sliced her corpse in two and made them into Sky and Earth, and from her spittle he made the clouds, wind and rain.

As with the other planets, our naming of Jupiter is from the Roman version of a Greek deity, in this case Zeus, chief of the gods. Early myths about Zeus run in striking parallel with aspects of Marduk, but in later Greek philosophical

Jupiter

thought, Zeus achieved the status of the abstracted essence of divine law. This philosophical elevation was already mythologically in place when he became the law-giver of the gods themselves, and the unrivalled father of the Olympians. The individual gods are unified in Zeus, who not only stands above the pantheon as supreme power, but appears as exponent of divine sway in general: it is to him that all prayers rise.

In traditional Western astrology, Jupiter is seen as one planet among a group of relative equals, although even then he never loses his imperial sway; he is known as the "greater benefic", forever seeking progress and expansion. He is broad-minded, exuberant, proud and imperious, and a generally honourable planet – in a way that recalls the mythological characteristics of the Roman god. Jupiter inclines to religion and philosophy, and is a wise counsel and teacher: the Indian name for the planet, Guru, emphasizes this role.

SATURN

Saturn

Second only to Jupiter in size, Saturn is 318 times the volume of Earth and more than 75,000 miles (120,700km) in diameter. At 886 million miles (1,426 million km) from the Sun, it is the farthest of the planets observed since ancient times, and fainter than Jupiter, Venus or Mars.

Since the first telescopic observations, this planet has impressed the popular imagination with its beautiful system of rings, composed of myriad tiny fragments orbiting like little moons along its equatorial plane. The system reaches an outer diameter of 170,000 miles (273,588km), but it is less than ten miles (16km) thick when seen edge on, and cannot be seen by the naked eye. Saturn's principal moon, Titan, which was discovered in 1665, is larger than Mercury and may be observed with binoculars; its chemically rich composition has excited scientific speculation that early conditions on the Earth may

Saturn

be replicated at the molecular level in this unlikely corner of the Solar System.

With an orbital period of 29.5 years, Saturn is the slowest of the ancient planets, and this seems to have led to two alternative strands of symbolic associations. In the medieval European vision of the universe, founded on the cosmology of Ptolemy, Saturn occupies the outermost planetary circuit, lying on the band beneath the sphere comprising the signs of the zodiac. Saturn was thought to be solitary, slow, cold and melancholy, and the ruler of time itself. In keeping with the sluggish nature assigned to the planet, it was identified with the metal lead. This is Saturn as marker of the outer limit of our reality. A related symbolism sees it as the enemy of life, the "greater malefic" – dirt, stagnant water and vermin all fall under its influence.

In the Chinese interpretation, however, Saturn is seen as representing

This illustration from an 18th-century Arabic manuscript shows different agricultural activities under the stewardship of Saturn.

Earth, the element at the centre of the Five Agents. The Chinese name for Saturn, "Quelling Star", has the connotation of exorcism of demons. On one point the traditions converge: the slow motion of Saturn gives it the character of age, and in Chinese astrology, as in Europe and Greek-influenced Asia, it is personified as an old man.

The origins of the Saturn of modern Western astrology are complex. In Mesopotamian myth, the planet belongs to the god Ninurta, the brother of Nergal (Mars). Ninurta rescued the tablets of fate from a dragon, and in gratitude, the gods granted Ninurta stewardship of the tablets. Thus he became the overseer of fate itself, prefiguring a significant thread in the later astrological Saturn.

The story of Cronos, the Greek Saturn, and his castration of his father Uranos (see page 50) places Saturn/Cronos first in the sequence of traditional planetary gods. In the next phase of the myth, the narrative repeats itself to produce the

Saturn

Saturn

second – and greatest – planetary god. As Uranos lay dying as a result of his castration, he and Gaia prophesied that Cronos too would be dethroned by one of his own sons. In order to prevent this fate, Cronos swallowed his children as soon as his wife Rhea bore them. The furious Rhea bore her third son, Zeus, in the dead of night and handed him into the care of Gaia; Rhea then gave Cronos a stone to swallow, which he thought was the infant. Zeus fulfilled the prophecy, overthrew his father and established his supremacy. Thus two successive ages of Saturn/Cronos and Jupiter/Zeus arose after the primordial reign of Gaia (Earth) and Uranos (Heaven), in which there was no time.

The development of the Greek Cronos into the Roman Saturn is at first puzzling, and introduces mythological elements that do not fit easily with the later, purely astrological Saturn. An Italian god of agriculture, Saturn was identified with a legendary early king of

A medieval illustration of Saturn (the Greek Cronos) about to devour one of his children. He wields the sickle with which he castrated his father. This implement may explain why the Romans equated Cronos with Saturn, a god of agriculture.

Saturn

Rome. His reign was so mild and civilized that it was later thought of as the Golden Age. The Saturnalia festival marked the winter solstice, when the Sun enters Saturn's sign of Capricorn. For the three days of the festival, public business was suspended, criminals were not punished, and slaves were allowed extraordinary licence. This popular event was extended to seven days under the later empire; and it is the pagan origin of our Christmas festivities.

The modern statistical research of Michel Gauquelin (see page 134) established an illuminating correspondence for Saturn. His research showed a significant occurrence of Saturn in the birthcharts of successful doctors and scientists. In astrology it is suggested that slow Saturn well reflects the disciplined, patient and cautious attitude required for success in the world of science.

THE MODERN PLANETS

The epithet "modern" refers to planets that were unknown in antiquity, their invisibility to the naked eye ensuring that they went unobserved before the era of modern technology. Astronomy made a great advance in the early 17th century with the invention of the telescope in the Netherlands. The heavens began to reveal wonders never before seen. By 1610, the Italian astronomer Galileo (1564–1642) was observing the moons of Jupiter using lenses that he had ground and polished himself, and which were capable of magnifications of up to thirty times.

It still took more than 170 years before the first of the modern planets, Uranus, yielded in 1781 to the telescope of British astronomer William Herschel (1738–1822), although over the previous century several observers had mistaken it for a star. In fact, Uranus can just be seen by the naked eye; but at twice the

The Modern Planets

A 4th-century CE silver plate portraying Neptune, the Roman god of the sea, surrounded by dolphins. The planet Neptune was discovered in 1846 by the German astronomer Johann Galle. It is said to rule the zodiacal sign of Pisces.

distance from the Sun as Saturn, it is faint and insignificant among many barely visible stars. Satellite probes have shown that Uranus is a surprising planet, with a system of rings and fifteen moons. Most strange is the 98° tilt on its axis of spin, giving it a motion quite different from any other planet.

Neptune was discovered in 1846. Its existence was first inferred from mathematical studies of unexplained perturbations on the orbits of Uranus and comets, indicating the presence of a

The Modern Planets

The Modern Planets

gravitational field. Neptune has approximately the same diameter as Uranus (33,000 miles/52,930km), but it is a massive 2,793 million miles (4,495 million km) from the Sun. It can be seen with binoculars, but only very faintly.

Pluto, the most recent planet to be discovered, is 3,670 million miles (5,906 million km) from the Sun. It was detected by photographic searches in 1930. With a diameter just a little greater than that of Mercury, this tiny planet can be seen with only powerful telescopes.

The modern planets raise intriguing issues, and serve as microcosms of how astrological and symbolic interpretations arise. How does a consistent and agreed symbolism emerge for these totally new celestial bodies, given the absence of an ancient tradition as a foundation? For the astrologer, operations such as naming a new planet are symbolically charged: contemporary events on the Earth must be mirrored in the "nature" of the new planet and vice versa. Therefore its

The Modern Planets

name must reflect this nature. When Uranus was discovered, Herschel wanted to name it "Georgium Sidus" after King George III of England, but this held little appeal for foreign astronomers. "Neptune" was proposed, along with unacceptable combinations such as "Neptune

An ancient Roman stone relief of Pluto (Hades), the god of the underworld, with his bride Proserpina (Persephone to the Greeks).

The Modern Planets

de George III". Eventually, the German astronomer Johann Bode suggested "Uranus", after the father of Saturn and grandfather of Jupiter. Most astronomers agreed that this was apt, although "Herschel" and "Herschellium" survived for many years, hence the "H" design of the planet's glyph (see page 8).

The symbolism of Uranus in modern astrology is largely independent of the myths of the god. However, in the case of both Neptune and Pluto, the myths colour the astrological interpretation. The connection is most obvious with Pluto, the Roman version of the Greek lord of the underworld, Hades. Astrological texts take the planet to symbolize "the underworld" of crime and depravity and also the deep unconscious.

Although astrologers place a great deal of importance on the name given to a new planet, the time of its discovery is a more enduring source of astrological interest. Uranus is linked with upheaval and revolution. Its discovery brought about the final upheaval of the secure medieval cosmos and threatened to sweep away the traditional symbolism of astrology and numerology, because instead of the sacred seven planets there were now an unholy eight. Around the same time, political revolutions raged in America (1775–83) and France (1789).

Neptune's historical associations are more complex. The first traces of modern spiritualism appeared in 1848, following supernatural occurrences at the home of the Fox sisters in New York State. The *Communist Manifesto* was published in the same year – indicative of Neptune's affinity with socialism, springing from compassion for the underclass.

Pluto's discovery is inevitably seen by astrologers as a herald of the cataclysmic events of the 1930s, including the rise of a terrifying new breed of dictator. It also shows, through the splitting of the atom, the release of new powers with the potential to annihilate.

The Modern Planets

A 16th-century map of Northern Hemisphere skies. The constellations' relative positions often reflect aspects of myth. Thus Ophiuchus (inverted, top right) represents Asclepius, who killed a scorpion that had stung Orion. His foot rests on Scorpio, as if to grind it into the ground as the two star-groups set.

CONSTELLATIONS

The constellations into which we group the stars depend solely upon our vantage point in the universe. The stars are myriad suns, at vastly different distances from us, but so remote that they appear as tiny twinkling lights, all equally far away. The modern constellations derive mostly from Greek interpretations of ancient Mesopotamian and Egyptian figures; but many other cultures have projected their myths into the stars. This chapter highlights stories and interpretations of some of the most important constellations.

This print of *c.*1510 shows the constellation Draco, the dragon, intertwined with Ursa Major and Ursa Minor. Draco was identified by the Greeks as Ladon, a monster slain by Herakles.

ORION

Mighty Orion, the Hunter, lays claim to being the most impressive of all the constellations, dominating winter nights in the Northern Hemisphere and the summer skies of the South. This figure, perhaps more than any other constellation, gives a sense of the connection with humanity that comes from contemplating the night sky. Countless cultures have imaginatively shaped the contours of this colossus striding high above the travails of human history.

Orion stands to the southeast of Taurus, with his belt of three stars slanted across the celestial equator. The second-magnitude star Mintaka (Delta Orionis), meaning "Belt", falls almost exactly on this great circle. A dagger hangs from his belt, its tip shown by Na'ir al Saif (Iota Orionis), the "Bright One in the Sword". With his right hand, on the eastern side of the figure, Orion brandishes a club. His right shoulder is

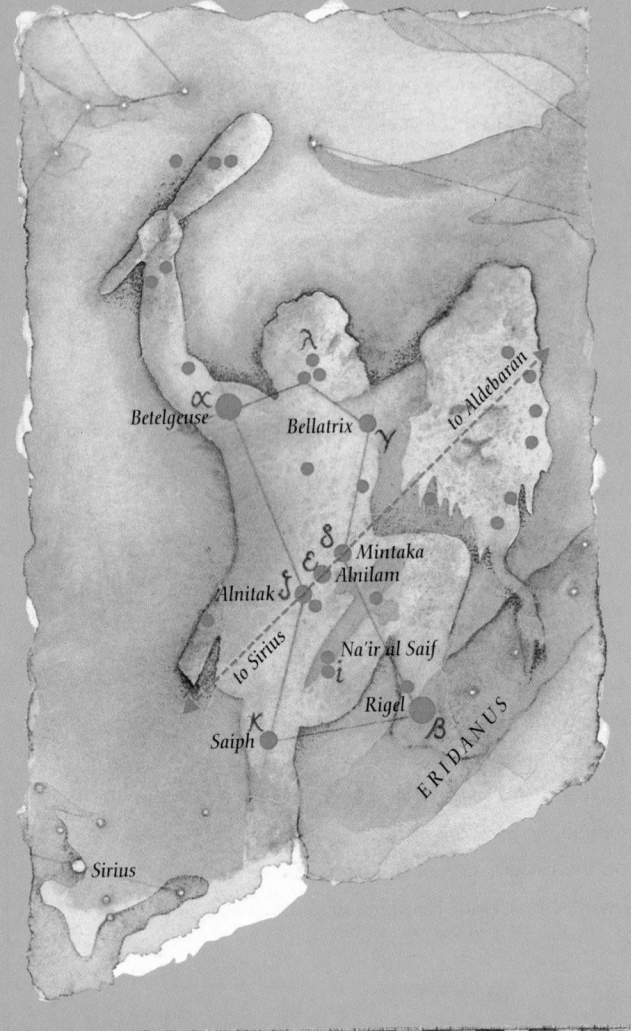

Orion, brandishing his club in his right hand and holding a lionskin in his left. To his northwest lies Taurus, with Sirius in Canis Major precisely to the southeast.

marked by the pale orange first-magnitude star Betelgeuse (Alpha Orionis); the left shoulder contains Bellatrix (Gamma Orionis), the "Warrior Woman", a pale second-magnitude star; and the giant's left leg is marked by Rigel (Beta Orionis), a blue-white star of the first magnitude. To the west of the figure a string of faint stars embodies a lionskin trophy held aloft by the hunter. The line of three stars of Orion's belt gives a useful orientation to other stars: extending it to the southeast brings us to Sirius in Canis Major; and to the northwest it leads to the star Aldebaran in Taurus.

As Orion sets, so Scorpio rises. This is reflected in the Greek myth that Orion was fatally wounded by a scorpion. Orion, a son of Poseidon, was a giant of great beauty and prowess. He lay with Eos (Aurora in Roman myth), the Dawn. She lingers in his presence, Orion fading slowly in the west as dawn breaks.

The constellation was revered by the earliest civilizations of the Euphrates

Orion

valley, but it is the Egyptians who have left us the clearest clues about the mystery of the stars of Orion, for them a representation of Osiris, god of the dead. Modern research on the Great Pyramid has demonstrated that its southern shaft, running from the King's Chamber at the core of the pyramid, was aligned to Orion's belt c.2700–2600BCE. The northern shaft was aligned to the Pole Star of that period, Thuban (Alpha Draconis). Fascinatingly, five of the Fourth Dynasty Giza pyramids repeat in their groundplan the pattern of the Orion constellation (see pages 290–294).

Osiris was the son of the Earth-god Geb and the Sky-goddess Nut. The first ruler on Earth, he took Isis, his sister, as consort. Osiris taught the people the arts of civilization and built the first temples. However, his jealous brother Seth mur-

DAS DAUS.

ORION.

Gebildet durch 63. Sterne deren
1. der ersten, 4. der andern, 9. der
dritten, 15. der vierten, 20. der fünften
17. der sechsten größe zubemercken.

Orion from a set of German playing cards of 1719. The figure is depicted in reverse, as if seen from outside the Celestial Sphere.

dered him and cut the body into pieces, which he scattered far and wide. The faithful Isis painstakingly reassembled his body and used her magic powers to revive him for long enough to conceive the god Horus. Osiris then descended to the underworld as eternal ruler of the dead. Protected by Isis, Horus grew to manhood to avenge Osiris and reclaim the throne.

In death, every pharaoh was identified with Osiris, while his successor was identified with Horus. It is now thought that the Great Pyramid (which in form probably represented the primeval mound on which the Sun was born) enabled the pharaoh to ascend to the heavens and unite with Osiris. After death, the pharaoh's embalmed body was placed in the King's Chamber, oriented with the passageway to Orion.

The Egyptians equated Orion with their god Osiris, shown here enthroned, with the four gods of the cardinal points.

© Orion

157

CANIS MAJOR AND MINOR

Canis Major and Minor

In Greek myth, no hunter would be complete without faithful hounds, and the powerful Orion is no exception. South of Gemini and east of Orion we see the giant's two hunting dogs, Sirius and Procyon. Sirius (Alpha Canis Majoris, magnitude -1.6) is the brightest star in our skies, far outshining a group of second- and third-magnitude stars which, with it, comprise the constellation Canis Major (the Greater Dog). Seen from the Northern Hemisphere, this dog sits low in the sky behind Orion, keeping a watchful eye on Lepus, the hare, a minor constellation at Orion's feet.

Procyon, a yellow-white star of the first magnitude, is roughly level with Orion's shoulders. It shares its small constellation, Canis Minor (the Lesser Dog), with just one other star of any significance, the third-magnitude Gomeisa.

The two dogs have a dramatic visual relationship with Orion in the

A pictorial star map showing the relative positions of Canis Major and Canis Minor. Like Sirius, Procyon is a binary (a double star); its companion-star is a white dwarf, first seen in 1896, known as Procyon B.

Canis Major and Minor

Northern Hemisphere's "Winter Triangle", a huge configuration straddling the Milky Way and formed by Sirius, Procyon and Betelgeuse (Alpha Orionis).

The name Procyon in the Greek means literally the "dog who rises before", because this star announces the coming of the more magnificent Sirius. The heliacal rising of Sirius (Sothis) – the day it first appeared as a morning star, when it would flash for a moment in the eastern sky just before the rising of the Sun – marked the start of the Egyptian year. This occurrence, in mid-July, heralded the onset of the annual Nile flood, on which the fertility of the land depended. A Sothic calendar was established by the middle of the 3rd millennium BCE. Sirius came to be identified with the powerful goddess Isis (see pages 156–157). Isis had marvellous powers of

The ancient Egyptian goddess Isis, wife of Osiris, was associated with Sirius. This illustration, after a mural in the temple of King Sety I at Abydos, shows her in the form Isis-Hathor.

Canis Major and Minor

magic: using earth fashioned from the spittle of the aged Sun-god Ra, she formed a snake which bit him. Ra could not comprehend the source of this potent venom and sought the healing power of Isis. Before she would lift the ailing god's affliction she made him reveal his true name to her; in so doing, he passed his power into her bosom.

The naming of Sirius as the Dog-Star also originates from Egypt. The "Dog Days" of folklore refer to the hot months of July and August, following the rising of Sirius, when the star's heat was believed to be added to the Sun's. In the Dog Days, it was said, vigorous plants and animals gained in strength, while the weaker creatures died.

The Fool, from a Tarot pack. Some authorities believe that it shows the dog Sirius at the heels of Orion (compare the illustration on page 154).

Canis Major and Minor

Sirius was known in China as Tian Lang, which means "Celestial Wolf". Associated with ill omens, a brightly showing Sirius was thought to portend attacks by thieves. It was also the governing star of Tibet.

In the 1940s it was discovered, amid some controversy, that the Dogon peoples of Mali in West Africa had traditionally used Po, a companion-star to Sirius with an elliptical 50-year orbit, as the basis for computing ritual periods. Yet astronomers had only established in 1862 that Sirius is indeed a binary (double star): the tiny Sirius B (magnitude 8.5) orbits it once every 50 years. How the Dogon could possibly have known this without the use of astronomical equipment remains an enigma.

GEMINI AND AURIGA

Culminating high overhead at midnight during January in the Northern Hemisphere, a pair of twins, represented by the first-magnitude stars Castor and Pollux (Alpha and Beta Geminorum), can be seen. This is the zodiacal constellation Gemini, the Twins. Castor, the more northerly twin, is white, while Pollux is orange. The constellation lies north of Procyon (Alpha Canis Minoris) and northeast of Orion. The feet of the Twins are marked by Alhena (Gamma Geminorum), between Al Nath (Beta Tauri) and Procyon.

In Greek myth, Castor and Polydeuces (Pollux in Roman myth) were referred to as the Dioscuri, "Sons of God": they were usually

A 16th-century Turkish representation of the astrological sign of Gemini. Through its planetary ruler Mercury, Gemini has an association with clerks and scholars.

Gemini and Auriga

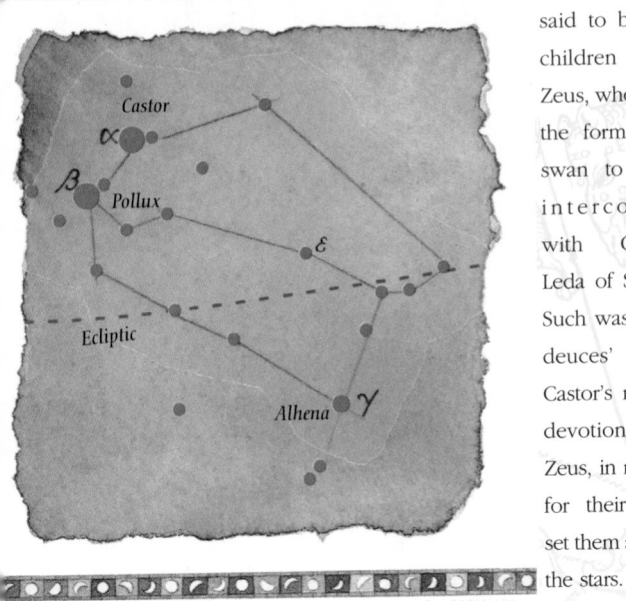

A star map showing Gemini. Johann Bayer, when assigning Greek letters in 1601, caused confusion by awarding beta to the brighter star, Pollux.

said to be the children of Zeus, who took the form of a swan to have intercourse with Queen Leda of Sparta. Such was Polydeuces' and Castor's mutual devotion that Zeus, in reward for their love, set them among the stars.

The seagod Poseidon gave the Twins the power to safeguard those at sea. In the Southern Hemisphere they can be seen high above the mast of the Argo, the ship in which Jason sailed to find the Golden Fleece (see pages 222–224).

The twins theme takes different forms across cultures. The Romans

associated this constellation with Romu-
lus and Remus, the descendants of
Aeneas and legendary founders of Rome
in 753BCE. In the Mayan creation myth,
they were a pair of peccaries copulating.
In early Phoenician and Chaldean tradition, they were thought to be a pair of
kids, following a shepherd represented
by Auriga, which lies west of Gemini
and north of Orion. The Arabs saw
them as peacocks, a designation which
survived into late medieval Europe.

Auriga, the Charioteer, carries two
kids in his hand and cradles a goat in his
left arm, marked by the magnificent yellow-white star Capella, the "Little She-
Goat" (Alpha Aurigae, magnitude 0.2).
The right foot of the charioteer passes
into the constellation Taurus, and is
shown by the tip of the Bull's horn, Al
Nath (Beta Tauri), lying just above the
ecliptic. As our gaze sweeps across the
sky, we pass through a spiral of stars
which crosses through Theta Aurigae
to the second-magnitude Menkalinam

Gemini and Auriga

Gemini and Auriga

(Beta Aurigae) near the right shoulder, and then across to Capella. Close to Capella the spiral turns through Epsilon Aurigae to mark the kids – the Hoedi, two minor stars (Nu and Zeta Aurigae). The figure has been perceived in its present form since the earliest records on the Euphrates.

In Greek myth, Capella was associated with the goat-nymph Amalthea, who suckled the infant Zeus. In return, Zeus turned one of her horns into the Cornucopia, the Horn of Plenty, which forever brimmed with food and drink.

Curiously, the charioteer has no chariot. Greek myth explains this by identifying him as Myrtilus, the servant of King Oenomaus. Myrtilus helped King Pelops to win the hand of Oenomaus's daughter Hippodameia. In return, Pelops had promised Myrtilus a night with his bride, but when Myrtilus claimed his prize, during a chariot ride on the wedding night, Pelops hurled him to his death in the sea.

Auriga, the Charioteer, shown holding the goat in his left arm, marked by the Alpha star Capella, and the kids in his left hand (the Hoedi: Nu and Zeta Aurigae). His foot is marked by Al Nath (Beta Tauri) and to his right lies the constellation Perseus.

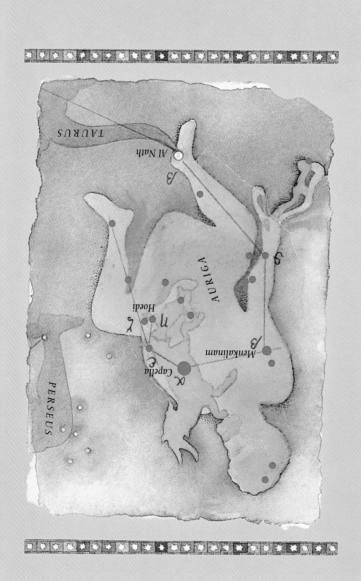

THE SOUTHERN SKIES

Once occupying a great span of the Southern skies, the constellation Argo is now split into three portions: Carina (the Keel), Puppis (the Stern) and Vela (the Sail). Only the flag on the stern is faintly visible from the Northern Hemisphere when it culminates in mid-January below Procyon. However, in the South, Argo forms a significant figure. The keel can be located from the beautiful brilliant-white first-magnitude Canopus (Alpha Carinae), south of Sirius.

In Greek myth, the Argo was the ship in which Jason and the Argonauts sailed to find the Golden Fleece (see pages 222–224). From ancient Egypt and Mesopotamia come stories of a celestial ship. The Greek historian Plutarch

The three constellations comprising Argo Navis (the Ship Argo): Vela the Sail, Puppis the Stern, and Carina the Keel. Here Canopus marks a rear oar.

claimed that it was the ship of the dead under the command of Osiris. Later Christian interpretations saw the vessel as Noah's Ark.

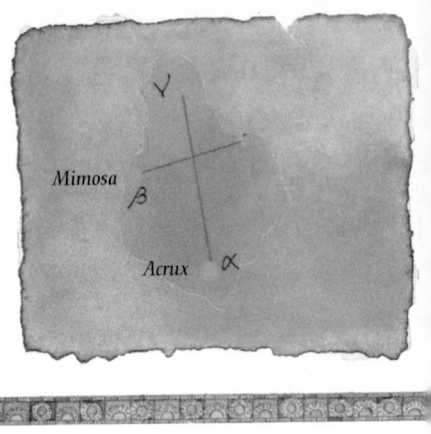

In some accounts, Canopus takes its name from the helmsman of King Menelaus of Sparta. For the Arabs its name, Suhail, was a synonym for wisdom. In China it was venerated by alchemists and known as the Old Man Star.

Another superb Southern Hemisphere constellation is Centaurus, the Centaur, culminating at midnight in April. Alpha Centauri, known as Toliman ("Shoot of the Vine"), or Rigel ("Foot of the Centaur"), is the closest star to our Sun, just 4.5 light years away. Because they are only 9° apart, Toliman and the beta star Agena have been treated as a pair. According to the scholar R.H. Allen, they were "Two Brothers" in aboriginal

A star map of the Southern Cross, whose long axis points toward the South Celestial Pole. It lies in a dense region of the Milky Way.

An ancient
constellation
map showing
Centaurus. He
appears holding
the small constel-
lation Bestia or
Lupus in his
right hand: the
two constella-
tions were often
depicted together

Australia and "Two Men who were
Lions" in South Africa. The centaur association originates in Greek myth. In one
account, Herakles accidentally shot the
wise and gentle centaur Chiron, who
had learned from Apollo the arts of
music, medicine, astrology and divination and was even credited with design-
ing the constellations.

Crux, the Southern Cross, once
formed the belly of Centaurus. Now it is a
distinct Southern circumpolar constellation, with four main stars in a compact formation. The vertical axis of the cross
points toward the South Celestial Pole.
The whole figure straddles the Milky Way,
4° wide at this point.
The Southern Cross was
just visible on the hori-
zon at Jerusalem at the time of
Christ, but has since slipped from
view, leaving the Northern Hemisphere, to quote Dante,
"famished and
widowed".

The Southern Skies

CANCER

An inconspicuous zodiacal constellation, Cancer the Crab lies to the east of Gemini. To find the constellation, one begins at Castor, the more northerly of the Gemini Twins, and constructs an equilateral triangle with a base line to Procyon – the apex of the triangle falls on the ecliptic at the Crab's centre on the fourth-magnitude South Asellus (Delta Cancri). Its fainter companion North Asellus (Gamma Cancri) lies 4° to the north. Between the two and slightly to the west is Cancer's distinctive feature, the star-cluster Praesape, which to the naked eye appears as a cloudy spot. There are more than 500 stars in this cluster, around eighty of which may be distinguished with good binoculars. To the south, two stars mark the Crab's legs: Acubens and, to its west, Al Tarf (Alpha and Beta Cancri).

Praesape, meaning "Swarm of Bees", seems an apt name for the star cluster, but it was also imagined as a manger, surrounded by asses – this is the literal meaning of Aselli.

The Greek myth of the Crab is as low-key as the constellation itself. It was crushed underfoot when it tried to nip the toes of Herakles as he fought the Hydra, a monster with a dog's body and eight or nine serpent-heads. The constellation Hydra lies just below Cancer.

In early Mesopotamian culture, however, Cancer was not so insignificant. It was the gateway through which souls passed from their sojourn in the stars down to their birth as human beings.

This 18th-century celestial map shows Cancer and surrounding constellations (clockwise from bottom left): Hydra, Ursa Major, Leo, Lynx, Gemini, Pegasus and Canis Minor.

Cancer

Cancer

The zodiacal sign Cancer has a resonance beyond the faintness of the stars in its namesake constellation, because the Sun's passage into the sign marks the northern midsummer solstice, when the Sun reaches its maximum height above the horizon and "stands still" (the literal meaning of "solstice"). This key moment in the solar calendar is marked by sight-lines at numerous Stone Age sites, such as Stonehenge (see pages 245–255). From the Earth, the latitude at which the Sun reaches its northernmost declination, at midsummer noon, is the Tropic of Cancer. The word "tropic" comes from the Greek *tropos*, "turning", and the solstices are, therefore, great turning points in the calendar, and in all human affairs.

Zosma

β

Denebola

ECLIPTIC

174

LEO

The impressive zodiacal constellation of Leo dominates spring nights in the Northern Hemisphere and autumn nights in the Southern Hemisphere. The regal lion pacing westward across the sky is easily identified, because Leo forms the first striking star-group east of Gemini and Procyon in Canis Minor. As our gaze passes over the inconspicuous

A star map of Leo. Cancer and Virgo flank the figure. The Sickle runs from Regulus (Alpha Leonis, just off the ecliptic) to Eta Leonis.

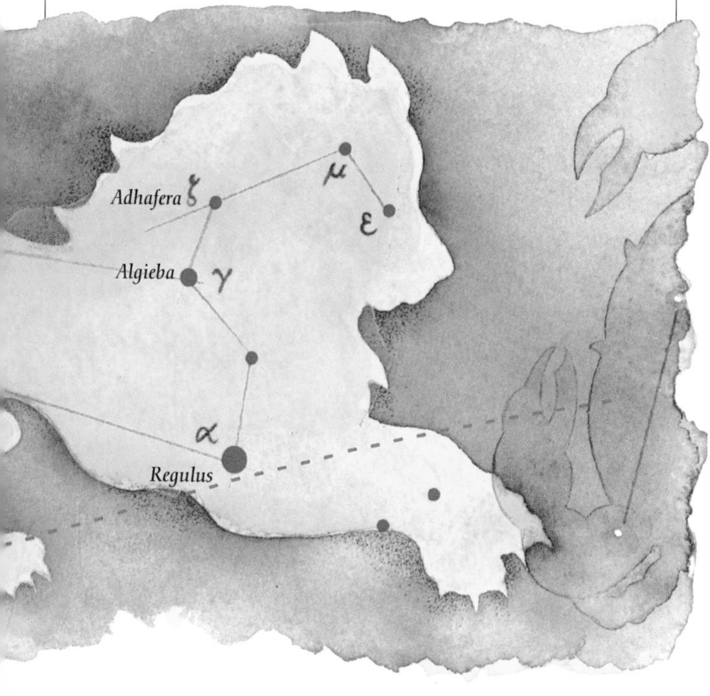

Adhafera η

Algieba γ

μ

ε

γ

α

Regulus

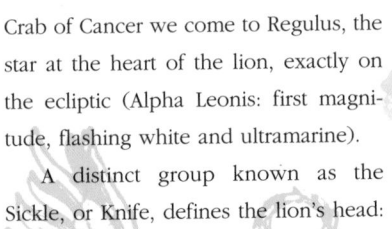

Crab of Cancer we come to Regulus, the star at the heart of the lion, exactly on the ecliptic (Alpha Leonis: first magnitude, flashing white and ultramarine).

A distinct group known as the Sickle, or Knife, defines the lion's head: these are the stars Algieba or "Lion's Mane" (Gamma Leonis), Adhafera (Zeta Leonis), with Eta and Kappa Leonis bringing us round to the nose. Regulus ("Little King"), now at the heart, was formerly part of the Sickle. Due west of the Lion's Mane, Zosma (Delta Leonis) marks the back and haunches, and Denebola is the "Lion's Tail" (Beta Leonis, second magnitude, blue).

Leo has held a position of eminence, and has been interpreted with a fair degree of consistency, from the earliest times and through all those cultures that can be seen as heirs to both Mesopotamian and Egyptian civilizations. This influence includes the Jewish, Greek, Latin, Indian, Persian and Arabic cultures, as well as later European astrology and mythology. Several inter-woven threads have contributed to the status of Leo as the "King of Signs, and Sign of Kings". One is undoubtedly the solar connection, making Leo a repre-sentative of the Sun. In the formative period of the settled civilizations of Mesopotamia and Egypt, some five mil-lennia ago, the Sun's passage at midday through this area of the sky coincided with the midsummer solstice. Leo was therefore the constellation of high sum-mer, which is manifestly the realm of the Sun. The Roman writer Pliny (1st century CE) records in his *Natural History* that the Egyptians worshipped Leo because the rise of the Nile coincided with the Sun's path through its stars. We have already seen the close association estab-lished by the Egyptians between Sirius (Sothis) and the Sun, because this star's heliacal rising in mid-July heralded the onset of the Nile floods (see page 160). The gates of canals irrigating the Nile valley were often decorated with a lion's

Leo

head, a possible origin of the motif of a fountain springing from the head of a lion, widely found in Greek and Roman architecture. It is unknown why a lion was chosen; but it is difficult to imagine a nobler animal and, once the connection of Sun and lion had been made, it seems to have become steadfast.

Regulus, sometimes called Cor Leonis, "Heart of the Lion", has come to take on all the magnificent associations of its constellation. However, this probably reflects a later tradition rather than the original Egyptian conception, since Leo was once a smaller constellation than our present figure, the stars of the Sickle being treated as a separate group. It appears that Mesopotamian astrology established Regulus as one of four "Royal Stars", guardians of the affairs of the heavens. These four stars are all of first magnitude and lie on or close by the ecliptic in a great cross: a line from Regulus in Leo to Fomalhaut in the opposite constellation of Aquarius forms one arm of the cross, while Aldebaran in Taurus and Antares in Scorpio complete the figure. In the first formative period of the major astral myths, these four watchers of the heavens, and their home constellations, marked the four stations of the solar year: the equinoxes and solstices.

Leo

In Greek myth, the constellation Leo was identified as the Nemean Lion, which Herakles skinned as the first of his twelve labours. The lion was an enormous beast, created by the Moon-goddess Selene, and its pelt was impervious to stone or metal. As it could not be defeated by weapons of any sort, Herakles had no choice but to wrestle with it. Although he lost a finger, he managed to strangle the creature to death. He skinned it by cutting the magical pelt with the beast's own claws. From then on, Herakles wore the pelt as invulnerable armour.

To the northeast of Denebola, the "Lion's Tail", above the figure of Virgo, lies a small cloud of minor stars which

An 1870 book illustration of the Leonids shower as observed by two French balloonists, Giffard and Fonvielle, while flying in their balloon *L'Hirondelle*.

forms the main group of the constellation Coma Berenices, or "Berenice's Hair", said to be the hair given by a queen to the gods in return for her husband's victorious homecoming. This group closely marks the North Pole of our galaxy, the axis around which our whole vast star system turns every 225 million years or so.

In the early morning around November 16 in the Northern skies, an occasionally impressive meteor shower, the Leonids, can be seen streaming from a point west of Adhafera in Leo. The phenomenon occurs as the Earth crosses the debris-strewn path of what is held to be a disintegrated comet. The Leonids are described scientifically as meteoric fireballs, or "bolides", which hiss as they cross the sky. Bolides were often thought of as flying dragons, and in the cultures of the Classical Mediterranean as well as in China they were regarded as messengers sent from the heavens.

VIRGO

Virgo, the Maiden, is a zodiacal constellation of great antiquity. In area it is the second largest star-group, although apart from its great star Spica (Alpha Virginis, magnitude 1.2, brilliant white), which lies just south of the ecliptic, the figure is not immediately obvious. It is most easily located by referring to the surrounding constellations – for example, if we continue the slight curve in the handle of the Plough just east of south to Arcturus (Alpha Boötis), and then follow the curve slightly westward for roughly the same distance, we come to Spica, the only other bright star in this region of the sky.

Virgo is shown as a maiden, frequently winged, lying along the ecliptic. In her right hand she holds a palm

Virgo

frond, and in her left a sheaf of corn: the "Virgin's Spike" of Spica.

Since the early Mesopotamian and Egyptian cultures, Virgo has been portrayed as a female figure associated with fertility. In Babylonian myth she represented Ishtar, Queen of the Stars. In the 8th century, the Venerable Bede erroneously linked Astarte, a goddess related to Ishtar, with Eostre, an old Anglo-Saxon goddess of the dawn and springtime: "Easter" derives from her name, which is actually related to "east".

Early Greek authors identified Virgo with the Egyptian goddess Isis, but principally the Greeks identified her as Persephone, the daughter of the Mother Earth-goddess Demeter. The connotations of fertility and harvest are represented in the myth of Persephone's abduction. She was spied among the springtime buds by Hades, the brother of Zeus and god of the underworld, who made a rare excursion into the upper world to seize the beautiful maiden and

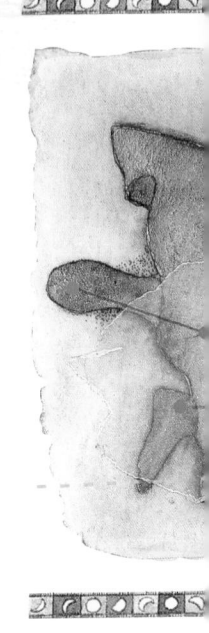

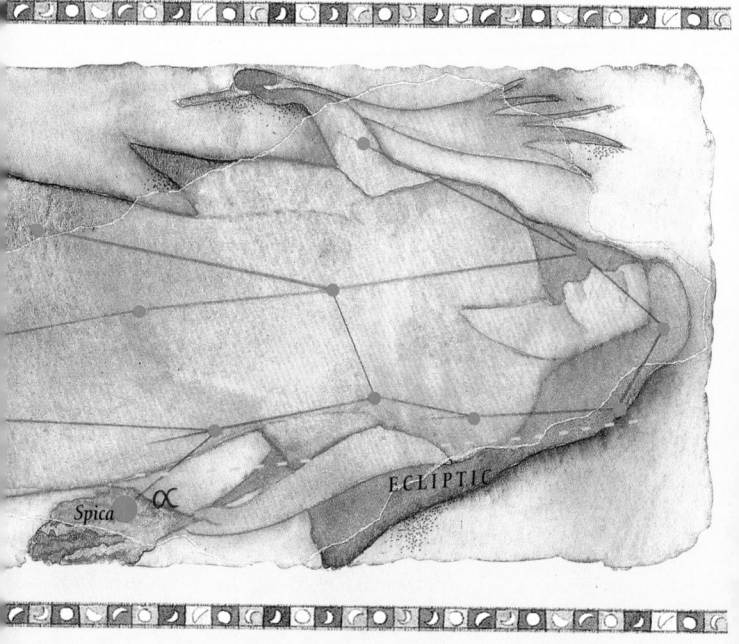

Spica ∝

ECLIPTIC

take her down into his realm to become his queen. Demeter spent many days wandering in search of Persephone until, in her rage, she threatened to withdraw her power of fertility and lay waste the Earth unless the gods restored her daughter. Zeus arranged for Hades to return Persephone, on condition that

A pictorial representation of the star-group Virgo. In her right hand she holds a palm frond; in her left, sheaves of corn marked by the bright star Spica (Alpha Virginis).

Virgo

the girl had not tasted the food of the underworld. However, she had already eaten some pomegranate seeds. Nevertheless, Zeus was filled with compassion for mother and daughter and decreed that thenceforth Persephone could spend half the year (spring and summer) above the Earth with Demeter and the other half (autumn and winter) below with Hades. When Persephone returns to her mother in springtime we see the renewal of fertility and growth.

In a second major strand alluded to by Classical authors, Virgo is Astraea, the "Star-maiden", the goddess of justice and daughter of Zeus and Themis. Astraea lived on Earth in the Golden Age and always urged people to judge others kindly. But humanity grew more sinful in the ensuing Silver and Bronze Ages, so she quit Earth for the heavens, where she may be seen as Virgo. Her scales of justice are the constellation Libra.

In Europe, Virgo has been identified since early Christian times with the

Virgin Mary; the brightest star, Spica, is said to represent the Divine Child cradled in his mother's arms. In Indian tradition, Virgo represents Kanya, the mother of the god Krishna, shown as a goddess seated before a fire.

Virgo, from *The Book of Fixed Stars* by the 10th-century CE Arab astronomer Al-Sufi, who was greatly influenced by Ptolemy. Virgo is seen as she would appear on the outside of the Celestial Sphere. In Classical representations, the figure usually has wings.

Virgo

Boötes and Corona Borealis

High in the Northern Hemisphere and a prominent constellation of the Northern spring and early summer, Boötes is easily identified by its principal star, Arcturus (Alpha Boötis, magnitude 0.2, golden yellow). This star is found by continuing the gentle curve of the handle of the Plough (Big Dipper), the Bear's tail of Ursa Major.

Arcturus marks the knee of the figure of Boötes ("Ox-Driver"), shown as a great hunter or herdsman with his head to the north and his feet bordering on Virgo. He is sometimes pictured with a pair of hounds on a leash to his west, represented by the small constellation of Canes Venatici ("Hunting Dogs").

Since Homeric times (*c.*8th century BCE), Boötes has been designated as an oxherd, but the name of his principal star, Arcturus, derives from the Greek for "bear-keeper". This reflects an alternative interpretation of Boötes in which he eternally drives the Greater and Lesser Bears around the Pole. Related names are the Waggoner, pulling the "waggon" of the Plough, and the Ploughman.

Another myth-complex identifies Boötes with Icarius, who was taught the secrets of wine-making by Dionysos (the Roman Bacchus). Icarius gave some wine to local shepherds, but they mistook its effects for poisoning and killed him. Led by his faithful dogs (Canes Venatici or Procyon in Canis Minor), his daughter Erigone found her father's body at the foot of a tree, where she hanged herself in grief. In memory of this tragedy, Icarius and Erigone were placed in the stars as the constellations of Boötes and Virgo (see pages 188–189).

The constellation Boötes, from Al-Sufi's *The Book of Fixed Stars*. As in the illustration from the same work on page 185, the figure is represented as if viewed from the outside of the Celestial Sphere.

Boötes and Corona Borealis

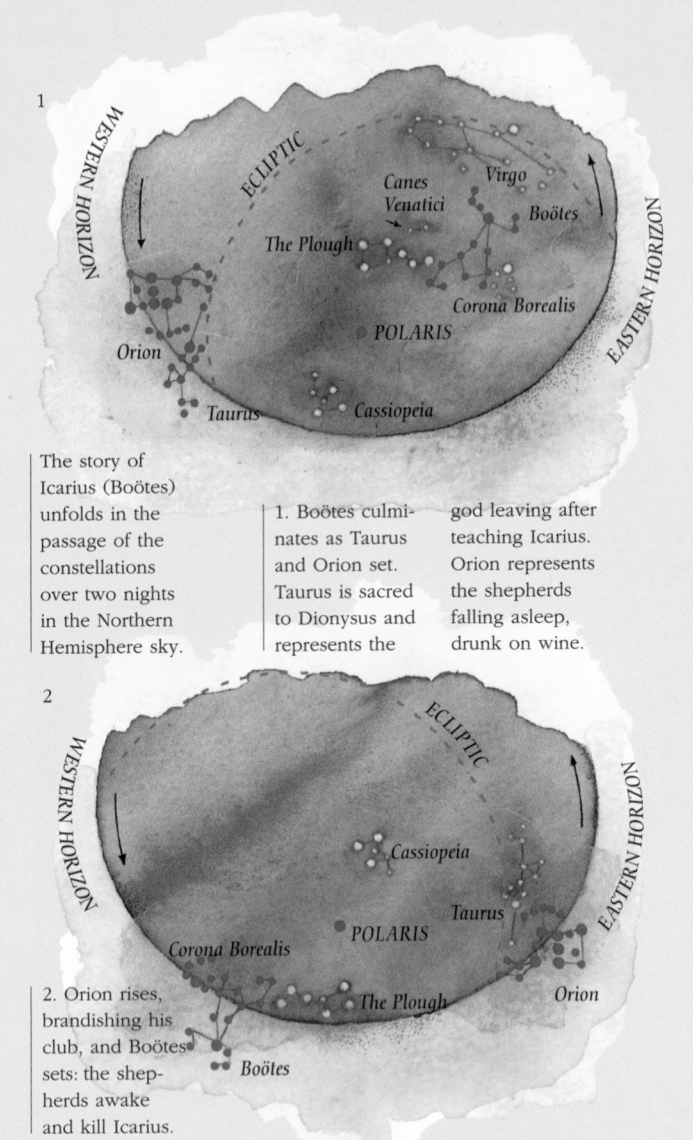

1

WESTERN HORIZON

ECLIPTIC

Canes
Venatici

Virgo

The Plough

Boötes

Orion

Corona Borealis

POLARIS

EASTERN HORIZON

Taurus

Cassiopeia

The story of Icarius (Boötes) unfolds in the passage of the constellations over two nights in the Northern Hemisphere sky.

1. Boötes culminates as Taurus and Orion set. Taurus is sacred to Dionysus and represents the

god leaving after teaching Icarius. Orion represents the shepherds falling asleep, drunk on wine.

2

WESTERN HORIZON

ECLIPTIC

Cassiopeia

Taurus

Corona Borealis

POLARIS

EASTERN HORIZON

The Plough

Orion

Boötes

2. Orion rises, brandishing his club, and Boötes sets: the shepherds awake and kill Icarius.

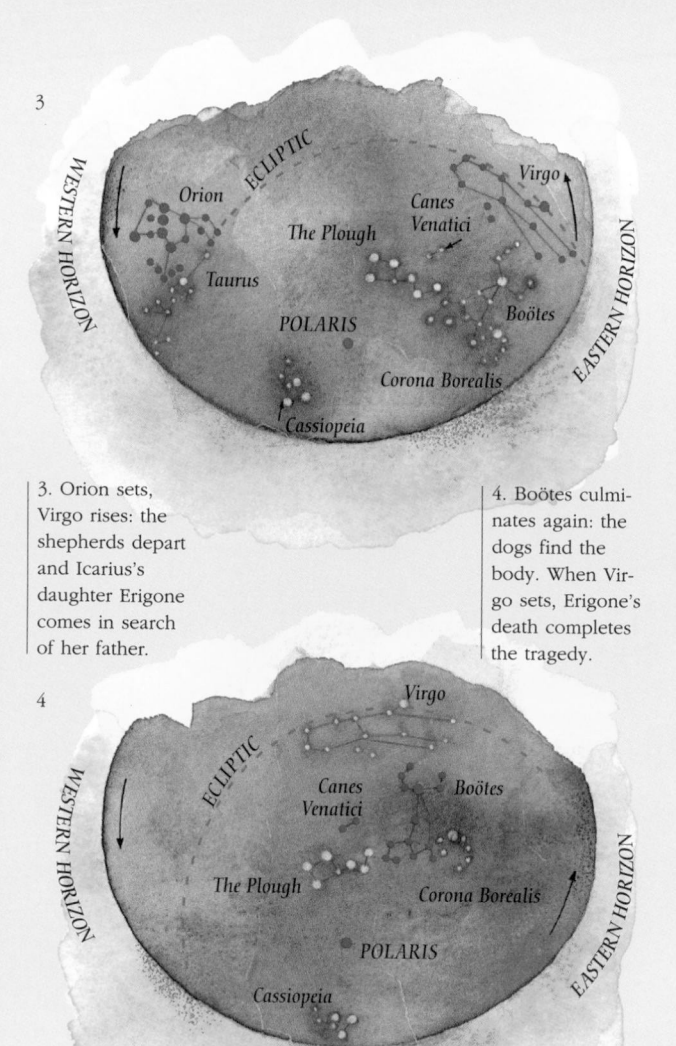

3

WESTERN HORIZON

ECLIPTIC

Orion

Virgo

Canes Venatici

The Plough

Taurus

Boötes

POLARIS

Corona Borealis

Cassiopeia

EASTERN HORIZON

3. Orion sets, Virgo rises: the shepherds depart and Icarius's daughter Erigone comes in search of her father.

4. Boötes culminates again: the dogs find the body. When Virgo sets, Erigone's death completes the tragedy.

4

WESTERN HORIZON

Virgo

ECLIPTIC

Canes Venatici

Boötes

The Plough

Corona Borealis

POLARIS

Cassiopeia

EASTERN HORIZON

Boötes and Corona Borealis

At the foot of Boötes is the small but distinctly beautiful constellation of Corona Borealis ("Northern Crown"). According to Greek myth, the crown is that of Ariadne, a princess of Crete, who gave a ball of twine to the hero Theseus so that he could find his way out of the dark Labyrinth after slaying the monstrous Minotaur. Ariadne left Crete with Theseus but he later abandoned her on the island of Naxos. However, the god Dionysos consoled her and married her. He flung her crown into the heavens, where it became the Northern Crown.

SCORPIO, OPHIUCHUS AND LIBRA

The distinctive hook of a scorpion's sting is clearly visible in the skies above the Mediterranean. This belongs to Scorpio, the eighth zodiacal constellation, on the portion of the ecliptic that lies to the south. Part of the tail is obscured at northern latitudes above 47°N and lost completely above 52°N; in Southern skies the whole constellation is visible. At the heart of the Scorpion is the ruddy-coloured Antares, the brightest star in this area of the sky and an ecliptic marker, lying less than 5° south of it. To the west of Antares, a fan of three stars forms the claws of the scorpion, which in early times extended to the faint stars that now form the scales of Libra. Scorpio can be found from Sagittarius, whose arrow points westward directly at Antares, just over 20° away.

About 5,000 years ago in Mesopotamia, the Sun's passage through Scorpio

brought the autumn equinox. Hence Antares was one of the four Royal Stars marking the equinoctial and solsticial points (see pages 178–179).

There is a malevolent, "death-bringing" theme associated with Scorpio, perhaps owing to the redness of Antares, which means "Rival of Ares". This link is amplified by the symbolism of the red planet, Mars (the Roman equivalent of Ares), known in astrology as a planet of slaughter, and the ruler of Scorpio.

In Greek myth, the stories of the scorpion, the hunter Orion and the healer Asclepius (identified with the constellation Ophiuchus) belong together. Orion once boasted that he would be able to kill all wild beasts. The Earth-goddess Gaia punished him for such arrogance, sending the scorpion to sting him on the heel. This is shown in the motions of the constellations. As Scorpio ascends on the eastern horizon, Orion dies and sets in the west. However, Asclepius healed Orion and crushed the

A star-map showing the relative positions of the constellations Scorpio, Ophiuchus and Libra. The brightest star in the group is Antares (Alpha Scorpionis).

Scorpio, Ophiuchus and Libra

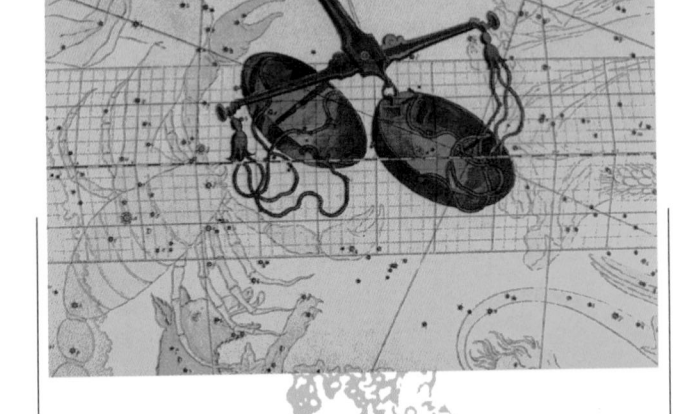

An 18th-century star map showing the constellation Libra.

scorpion underfoot – reflected by Orion rising again in the east, restored to vigour, as Asclepius (Ophiuchus) stamps on Scorpio in the west.

The name Ophiuchus derives from the Greek for "Serpent-Handler". There was no hero or god of this name, but the figure is associated with the healer Asclepius (Asklepios, the Roman Aesculapius), son of Apollo, whose emblem – shared with Hermes/Mercury – of serpents entwined around a staff has long been a symbol of the medical profession. Ophiuchus stands over the scorpion with his head to the north, beneath the inverted figure of Hercules, so that the heads of these heroes almost touch. In his hands he grasps a serpent, the constellation Serpens, which was originally one constellation but is now separated by Ophiuchus into Serpens Caput ("Serpent Head") and Serpens Cauda ("Serpent Tail"). There is no bright star in the group. The main star, in the head of Ophiuchus, is Ras Alhague (second magnitude), Arabic for "Head of the Serpent-Charmer".

To the east of Ophiuchus is the small and visually insignificant zodiacal constellation of Libra. It can be located on the ecliptic roughly midway between Spica in Virgo and Antares in Scorpio. Near the mid-point is the fulcrum of the scales, the star Zuben Algenubi (Beta Librae), almost exactly on the ecliptic. Around 8° northeast is the slightly brighter Zuben Elschemali.

The gorgon Medusa's head: a roof ornament from the temple of Apollo at Delphi. After Perseus cut off Medusa's head, Athene gave two phials of blood from the wound to Asclepius (Ophiuchus). Using blood from the left side, he raised the dead; with blood from the right he could kill instantly.

Scorpio, Ophiuchus and Libra

THE HERCULES GROUP

Hercules is a large and somewhat indistinct constellation rising high in the Northern midsummer sky. None of its stars are much above the third magnitude in brightness. For the Northern Hemisphere observer, the figure appears inverted, with his feet toward the North Pole. He is located to the west of the star Vega (Alpha Lyrae), and above Ophiuchus. The brightest star, which marks Hercules' head (Ras Algethi, "The Kneeling Man's Head"), is only a few degrees to the west and a little north of the more noticeable Ras Alhague, in the head of Ophiuchus. Hercules rests his foot on the head of the dragon Draco, which coils round the North Celestial Pole.

This constellation's undramatic appearance hardly seems to do justice to the most famous of all Classical heroes, Herakles (Hercules to the Romans). Most Greek authors knew the figure as "The Kneeling One", without under-

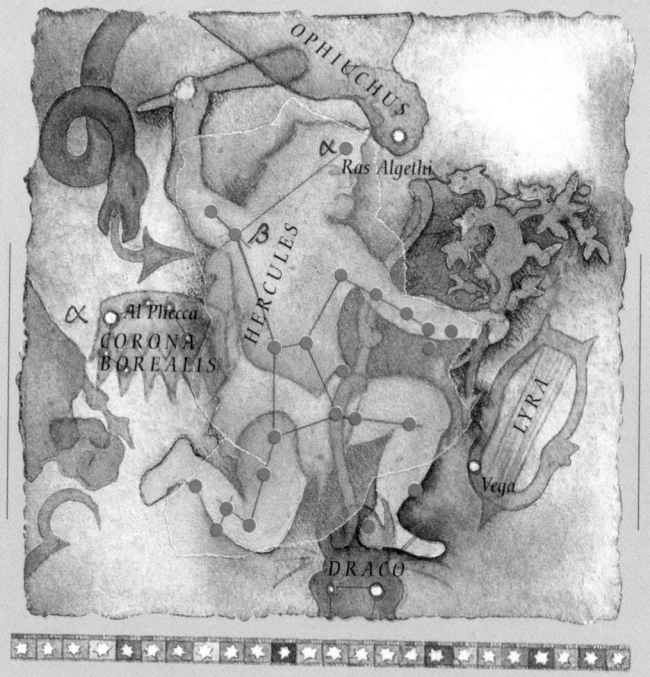

Star labels on the map: OPHIUCHUS, α Ras Algethi, β, HERCULES, α Al Phecca, CORONA BOREALIS, LYRA, Vega, DRACO

standing its derivation. However, Herakles is a close equivalent to the Babylonian hero Gilgamesh: among a number of parallels, both defeat a monstrous lion, a divine bull and a dragon. From the late 4th millennium BCE there is a Mesopotamian description of a hero, later identi-

A star map of Hercules and neighbouring constellations. Hercules grasps the Hydra, a monster he slew as one of his twelve labours.

fied with Gilgamesh, kneeling with one foot on a dragon's head – an apparent prototype of the Greek "Kneeling Man".

The Romans finally established the tradition of identifying the constellation with Hercules. The sources vary in the details of the story, but agree on its broad outline: the hero was required to undertake twelve seemingly impossible tasks as penance for killing his children in a fit of madness. In the night sky he is imagined wearing the indestructible pelt of the Nemean Lion (associated with the constellation Leo), which he killed as the first of his twelve labours.

One labour, the cleaning of the Augean stables, has become a well-known literary metaphor to express the idea of eradicating corruption and rubbish. King Augeas owned 3,000 oxen, but their stalls had not been cleaned for thirty years. Herakles had the filthy task of clearing them out in one day, which he achieved by diverting two rivers through the stables.

The Latin writer Servius (*c*.300CE) commented on the tradition of attempting to interpret the twelve labours of Herakles as the twelve signs of the zodiac, with the hero as a Sun-god passing on his journey through the year. But the interpretation has a serious flaw: the early Greeks gave fewer than twelve labours, and the labours as they later came to be identified are not ordered in the sequence of the zodiacal signs.

Lying east of Hercules on the Milky Way is a group of relatively small constellations, of which three (Lyra, Aquila and Cygnus) are visually striking by virtue of their bright stars. Lyra, the Lyre, close by Hercules, is marked out by the beautiful pale sapphire Vega. With a magnitude of 0.14, it is the fourth

The Hercules Group

An 8th-century BCE Babylonian relief of Gilgamesh, who was later associated with Herakles.

The Hercules Group

brightest star in the heavens. The Greeks called this constellation Chelys, the Tortoise-shell, from which Hermes made a marvellous lyre for the musician Orpheus.

Southeast of Vega and across the eastern edge of the Milky Way is another brilliant star, Altair (magnitude 0.9, pale yellow), in the constellation Aquila, the Eagle. This royal bird served Zeus loyally. Its final mission was to punish the Titan Prometheus for stealing fire from the gods and giving it to humankind. Prometheus was chained to a mountain peak and every day the eagle devoured his liver, which each night was miraculously restored. Herakles eventually slew the eagle, and Zeus placed it among the stars.

SAGITTARIUS AND CAPRICORN

Lying well south of the equator, Sagittarius is not a prominent constellation for Northern Hemisphere viewing, and at middle latitudes it clings just above the horizon in the light summer nights of June to early August, never completely revealed. But at Southern latitudes during these months it can be seen prominently, high in the sky.

The ninth zodiacal constellation, Sagittarius, the Archer, is represented as a centaur, half-man and half-horse. He is armed with a bow and arrow, forming the western part of the figure, which falls on the Milky Way, a wide band at this point.

The constellation Sagittarius, showing the major stars and, particularly, the formation of the arrow that points toward Antares in Scorpio.

ECLIPTIC

Kaus Borealis
Nunki

Kaus Medius

δ
to Antares

Al Nasl
γ

ε
Kaus Australis

α Rukbat

β

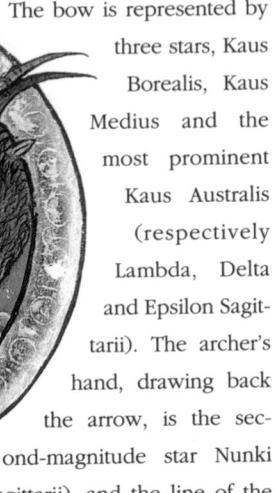

The bow is represented by three stars, Kaus Borealis, Kaus Medius and the most prominent Kaus Australis (respectively Lambda, Delta and Epsilon Sagittarii). The archer's hand, drawing back the arrow, is the second-magnitude star Nunki (Sigma Sagittarii), and the line of the arrow stretches from Kaus Medius to Al Nasl (Gamma Sagittarii), which marks its point. The arrow appears to be aimed at Antares in Scorpio, some 20° westward on the farther edge of the Milky Way.

Historically, the representation of Sagittarius as a centaur has led to confusion with the Southern constellation Centaurus. However, each has a distinctive personality; unlike the mild Southern centaur, Sagittarius is wild and war-

The zodiacal sign Capricorn, depicted as an amphibious creature with the head of a goat and the tail of a fish, from an English Psalter of *c*.1170.

like: he can be traced to the Mesopotamian archer-god Nergal, who ruled
the planet Mars. In Greek myth, however,
Sagittarius was identified with the wise
centaur Chiron. It was said that the goddess Artemis caused the death of Orion
by arranging for a scorpion to sting him
on the heel. Chiron killed the scorpion
with an arrow, shown in the skies by the
centaur aiming at the scorpion's heart,
Antares. This story overlaps that of the
crushing of the scorpion by Asclepius
(Ophiuchus; see pages 193–195).

The tenth zodiacal constellation,
Capricorn the Goat, is an obscure figure
for Northern Hemisphere observation,
lying well south of the equator and with
no bright stars. It is best observed when
it culminates around midnight during
August, and may be found by drawing a
line from Vega in Lyra through Altair in
Aquila; this runs on to define the horns
and the head of the goat through the
figure's alpha and beta stars, Giedi and
Dabih. The creature's fish-tail is marked

Sagittarius and Capricorn

Sagittarius and Capricorn

by the gamma and delta stars Nashira and Deneb Algedi ("Goat's Tail").

Classical authors often treat Capricorn simply as a goat; the goat-fish image can be traced to Mesopotamian origins. There is evidence to connect the figure with the Chaldean deity Oannes, the god of wisdom, who was half-man, half-fish. In India, Capricorn was variously shown as an antelope or a crocodile, and occasionally as a hippopotamus with a goat's head – associations that repeat the theme of a creature that moves between water and land.

A star map of Capricorn. The alpha star marks the horns and the delta star marks the tail. Capricorn is the smallest of the twelve zodiacal constellations.

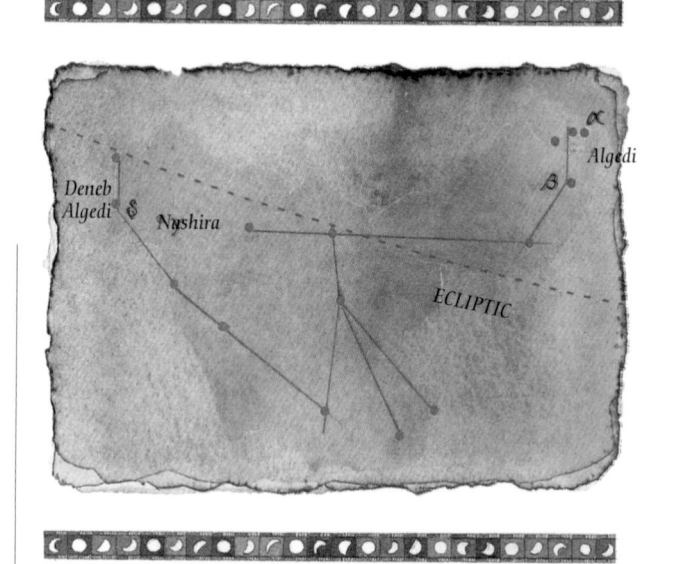

AQUARIUS AND PISCES

The eleventh zodiacal constellation, Aquarius, the Water Carrier, culminates at midnight from August to September. The main part of the figure lies south-west of the great Square of Pegasus. He is shown as a man pouring a pitcher of water, from which flows a gushing stream known as Fluvius Aquarii, the River of Aquarius, which curves around by the figure's feet. Apart from this star Aquarius is not easily discernible, its stars being of the third magnitude or fainter. But the constellation nevertheless has a fascinating and consistent history.

Our modern version of the constellation is almost identical to a water carrier found in early Babylonian carvings, then as now awkwardly and unaccountably reaching back with his free arm toward Capricorn. There is a strong likelihood that the figure is associated with the myth of the great flood recorded in sources from the 2nd millennium BCE.

Aquarius and Pisces

Aquarius and Pisces

Furthermore, in this period the 11th month, corresponding to the Aquarius period of January–February, was termed the "curse of rain".

The Greek poet Pindar (*c.*522–*c.*422BCE) recorded the ancient belief that the constellation symbolized the spirit of the source of the life-giving Nile. The ancient Egyptians deified the river as the god Hapi, depicted residing by the primordial spring from which he distributed water to heaven and the Earth from his urns. In Greek myth, Aquarius represents Ganymede, the beautiful Phrygian boy ravished by Zeus in the form of an eagle. He became the cup-bearer of the gods.

The humane symbolism of Aquarius has come into the forefront of the modern imagination owing to the widespread "New Age" culture for which it stands – an idea that has become inextricably bound up with the turn of a new millennium. The slow effect of precession (see pages 40 and 42) means that

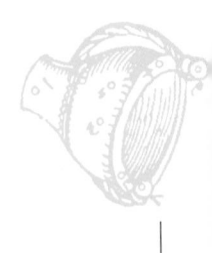

A star map showing Aquarius and Pisces among Pegasus, the sea-monster Cetus and Capricorn.

Aquarius and Pisces

another three centuries will pass before the spring equinox comes into alignment with the easternmost part of Aquarius. Whatever significance we give to this, it has come to represent hope for better things.

The neighbour of Aquarius is Pisces, the Fishes, the twelfth zodiacal constellation. It is a diffuse spread of faint stars covering a large span of sky between Pegasus and Aries. The fishes are joined at their tails by a cord, knotted at their main star, Al Rischa (Alpha Piscium). One fish appears to strive vertically northward, whereas the other stretches along the ecliptic, below the Square of Pegasus. Pisces culminates at midnight during the period from September to October. This constellation has been shown as fishes since Babylonian times, although its form has varied.

The 11th-century Arab astronomer and astrologer Al-Biruni was almost certainly correct to assert that the original star-group consisted of one fish. The

Greek scientist Eratosthenes (born 276BCE) traced it to a Syrian goddess, Derke (Atargatis to the Greeks), a huge fish with a woman's head. The Greeks compounded Atargatis with the Syrian deity Astarte, who was in turn associated with the goddess Aphrodite. In one account, a monster startled Aphrodite and her son Eros (Cupid to the Romans), who escaped by turning into fishes and leaping into the sea. To avoid losing each other, they tied their tails together. In the Roman version, two fishes carry Venus and Cupid to safety.

A 16th-century Turkish painting of Pisces, here depicted as a single fish.

Aquarius and Pisces

THE ANDROMEDA GROUP

The story of Andromeda, and her rescue
from a sea monster by Perseus mounted
on the winged horse Pegasus, is one of
the best known of all the Greek myth-

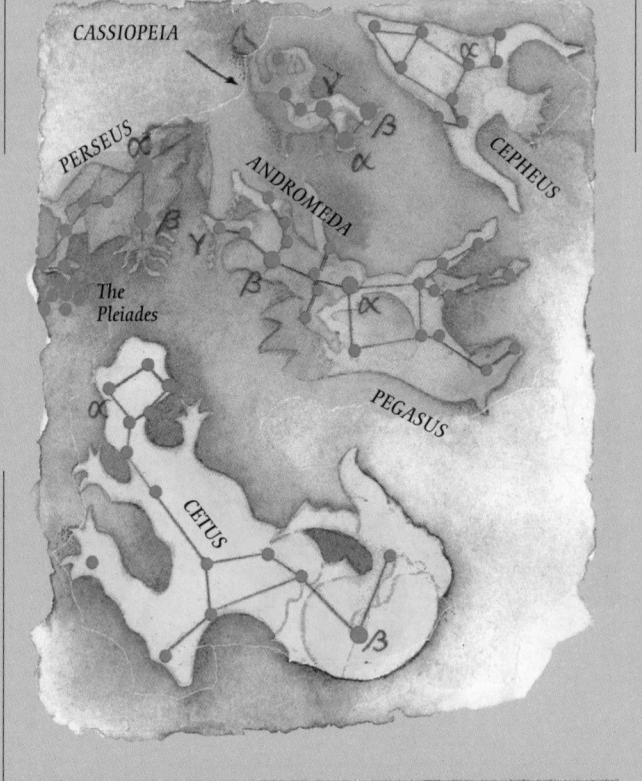

The Andromeda Group

complexes, and every autumn a striking, and easily located, group of constellations unfolds the drama across a large sweep of the Northern sky.

The starting point for observation is the circumpolar constellation Cassiopeia, immediately identifiable as a letter W (or M when inverted) formed from an arrangement of five stars of the second and third magnitudes. It is a signpost to the Celestial Pole, as the W is cupped toward Polaris, the Pole Star. Also, the configuration lies the same distance (30° of declination) on one side of the Pole as the Plough on the other side. By locating either or both of these figures, therefore, the whole Northern sky can be oriented at a glance.

Moreover, in our epoch these constellations allow a ready identification of the equinoctial colure, the meridian line running through the Poles to the spring and autumn equinox points. The colure skirts the edge of the W at Caph (Beta Cassiopeiae) and runs on down to the

A star map showing the relative positions of Perseus, Andromeda, Cassiopeia, Cepheus and Pegasus, along with the terrible sea monster Cetus, which threatened the life of Andromeda before being slain by Perseus.

Northern spring point. In the other direction, the colure passes through the Pole and then down through the Plough between Phecda and Megrez (Gamma and Delta Ursae Majoris), which mark the "handle" of the Plough.

The next stage of the visual journey is the "Square of Pegasus", a beautiful group of four stars culminating at midnight during September. The brightest of the four lies just 2° east of the equinoctial colure and due south of Caph. This is the second-magnitude Sirrah ("The Navel of the Horse") or Alpheratz; it has in modern times become the alpha star of the constellation Andromeda, and marks the head of the chained princess, close by the winged horse.

The northern edge of the Square of Pegasus extends into a gentle east-to-northeast curve of four stars describing the figure of Andromeda, from Sirrah at the head through Mirach at her waist to Alamak marking her feet. At her feet and to her east is her saviour and suitor

A view, from a work by Tycho Brahe (1602), of the constellation Cassiopeia that shows the 1572 Nova (to the right), a faint star that explodes and shines brightly before decreasing to its normal luminosity.

The Andromeda Group

212

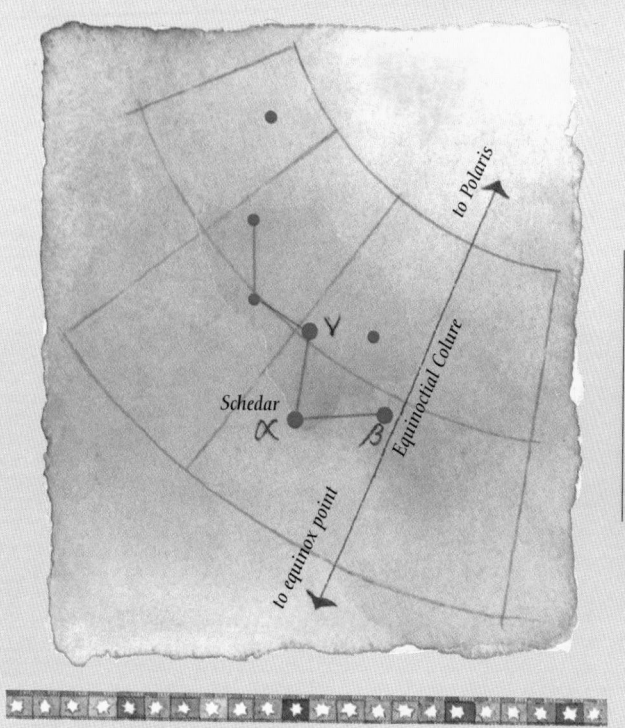

In the image: *to Polaris*, *Equinoctial Colure*, *Schedar*, α, β, γ, *to equinox point*

Perseus, located above the Pleiades in Taurus. His head is marked by the second-magnitude lilac-coloured Mirfak or Algenib (Alpha Persei). In his left hand Perseus holds the Medusa's severed

A star map of Cassiopeia, defining the distinctive "W" shape of the main stars.

A star map of Pegasus, the mythological winged horse. Clearly seen is the great "Square of Pegasus", in the top left hand corner of which sits the star Sirrah, which is known as either Alpha Andromedae or Delta Pegasi.

head, shown by ill-omened Algol (Beta Persei), the demon star; this is a white eclipsing binary (double-star) system which "blinks" by varying its brightness between the second and third magnitudes in a cycle of two days 21 hours.

Two other characters in the drama remain to be located. Taking a westward line from Mirfak in the head of Perseus and back through Cassiopeia, we come to her circumpolar husband, Cepheus;

A pictorial view of Andromeda as seen in the Northern Hemisphere. Alamak, the gamma star, is in fact a triple star, including one blue and one orange star.

and far away to the south, beneath Perseus and Aries, lies the sea monster Cetus, his snout marked by the third-magnitude, orange star Menkar (Alpha Ceti), just above the celestial equator.

The story of Perseus repeats a recurring motif in Greek myth, that of the son who is prophesied to kill one of his pro-

The Andromeda Group

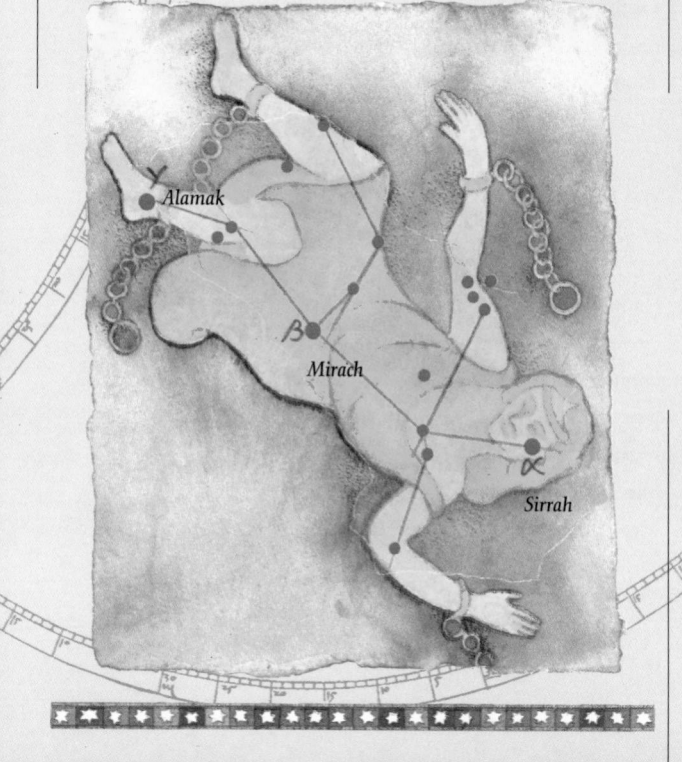

Alamak

β

Mirach

α

Sirrah

The Andromeda Group

genitors – in this case, his grandfather. An oracle foretold that King Acrisius would have no male heirs, but that his grandson would kill him. To circumvent this doom the king had his daughter Danaë imprisoned in a brass tower. However, Zeus entered her chamber disguised as a shower of gold, and in this form he impregnated her. Danaë managed to conceal her child, Perseus, for four years, but when Acrisius finally discovered his grandson, he had mother and child locked in a trunk and cast out to sea. However, the trunk was carried to the island of Seriphos, where Danaë and Perseus were rescued by the brother of Polydectes, the island's king.

As a young man, Perseus defended his mother against the advances of Polydectes. The king vowed to choose another bride if Perseus agreed to bring him the head of the Gorgon Medusa. Around the head of this terrifying creature coiled venomous snakes, and all who gazed on her turned to stone.

However, the gods chose to help Perseus. Athene gave him a polished shield in which he would be able to see Medusa's reflection without looking directly at her, and Hermes provided a sickle to slay her. He also acquired winged sandals for flight, a helmet of invisibility from Hades, and a pouch in which to carry Medusa's head.

Perseus flew to the land of the Gorgons, who were asleep among the petrified forms of beasts and men. Using the reflection in his shield, he struck off Medusa's head with one blow. From her body sprang a warrior and the winged horse Pegasus. Perseus, invisible, flew away with the head in his pouch.

On his return journey he came upon Andromeda chained to a rock. She was the daughter of King Cepheus and Queen Cassiopeia of Ethiopia. Cassiopeia once foolishly compared her and Andromeda's beauty to that of the Nereids (sea nymphs). They complained to Poseidon, who summoned a flood and

The Andromeda Group

The Andromeda Group

created Cetus, a sea monster, to ravage the people. An oracle told Cepheus that he could save his kingdom only by sacrificing his daughter, so he had Andromeda chained naked to a rock at the sea's edge as an offering to the monster.

Perseus, chancing on the scene, was entranced by the beauty of the girl. Cepheus and Cassiopeia agreed to let him marry her if he could save her. Dressed in his cloak of invisibility, Perseus swept down from the air and beheaded the beast with the sickle.

Perseus hurried back to Seriphos, but Polydectes had reneged on his promise not to pursue Danaë. Perseus stormed into a banquet held by Polydectes and pulled out the Gorgon's head, instantly turning Polydectes and his minions into a circle of boulders.

Perseus gave the head of Medusa to Athene – who thereafter displayed it on her shield – and returned to Argos with his mother and bride. Acrisius, remembering the original oracle, fled to

A pictorial representation of the constellation Perseus. The star Mirfak (Alpha Persei) marks his head and Algol (Beta Persei) denotes the head of the gorgon Medusa in his left hand. His winged sandals are marked by the gamma star at his feet.

Larissa in Thessaly. One day Perseus was invited to take part in funeral games at Larissa, an occasion that Acrisius attended. When Perseus threw the

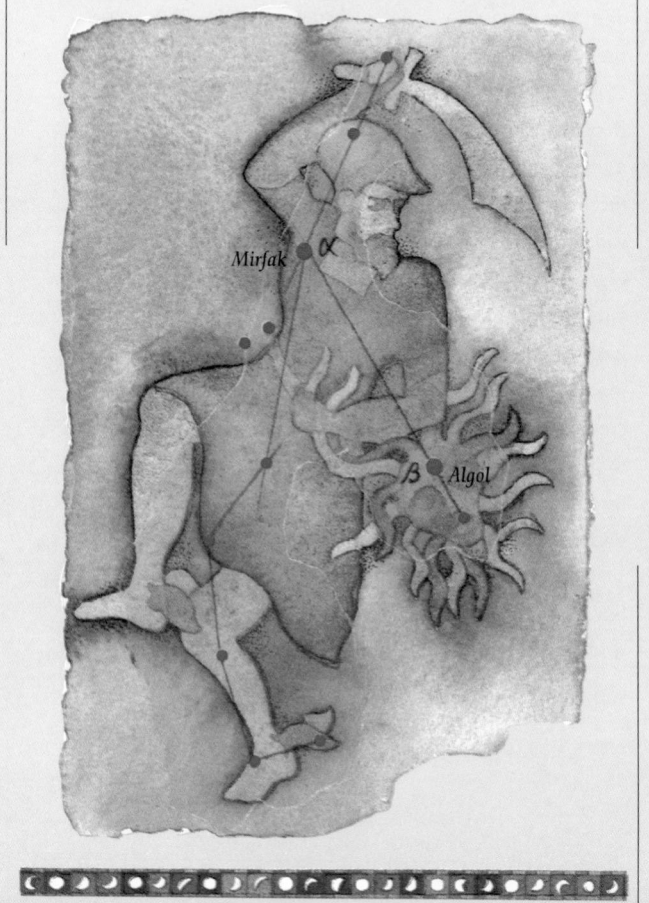

The Andromeda Group

The Andromeda Group

discus, the gods blew it off course so that it struck Acrisius and killed him.

Pegasus presents an intriguing set of mythological motifs. The Greeks did not originally represent Pegasus as winged, but abundant evidence from Etruscan and Mesopotamian sources shows that the winged horse was known in pre-Classical times. It was commonly depicted as the steed of a great hero or a god. In ancient Israelite legend he was the horse of the legendary warrior Nimrod, "the mighty hunter before the Lord" (Genesis 10.9). In the earliest stratum of Greek myth, Pegasus is the bearer of Zeus' thunder and lightning. By stamping his hoof on Mount Helicon, Pegasus created the Hippocrene, the "Fountain of the Horse" – a fountain of poetic inspiration sacred to the Muses, the nine goddesses of the arts and sciences.

Some versions of the Perseus story have him riding Pegasus to save Andromeda, but the principal Greek myth relating to Pegasus is the myth of

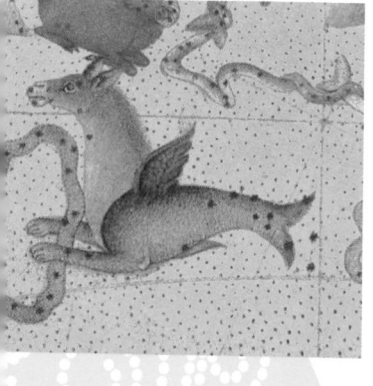

The constellation of Cetus, depicted in an Indian star map (see page 218).

Bellerophon, son of the king of Corinth. Bellerophon was set the seemingly impossible task of slaying the Chimera, a monster with the fire-belching head of a lion, a goat's body and a serpent's tail. He was advised by a seer to catch and tame Pegasus, which he did with the aid of a golden bridle given by Athene. Borne aloft by Pegasus, Bellerophon attacked the Chimera and thrust between her jaws a lump of lead attached to the end of his spear. The monster's breath melted the lead, which ran down her throat and destroyed her inner organs.

One version recounts that Bellerophon later presumed to ride Pegasus to Mount Olympus, the home of the gods. Zeus sent a gadfly to sting Pegasus, which reared, flinging Bellerophon to Earth. From that day, he wandered blind and lame, shunning human contact. Pegasus became a lowly pack-horse.

The Andromeda Group

ARIES

Aries, the Ram, culminates at midnight during October. It is located to the west of the Pleiades and Taurus and to the southwest of Perseus. Its designation as the Ram goes back to Mesopotamia of the 3rd millennium BCE. However, the shape of the ram in repose on the line of the ecliptic is difficult to distinguish apart from the dominant clump of three main stars that define the head. The alpha star Hamal (magnitude 2.2, yellow), is the "Horn Star" or "Ram's Eye".

The first zodiacal constellation, Aries marked the spring equinox (the crossing of the First Point of Aries with the celestial equator; see page 42) for two millennia before our era. Various temples dating from the middle of the 2nd millennium BCE have been found oriented to Hamal.

In Greek myth this constellation represented the Golden Fleece shorn from

Aries

A star map of Aries, the Ram, the first sign of the zodiac. It lies just above the ecliptic, and in visual representations its largest star, Hamal, usually marks either the ram's eye or its horn.

a magical flying ram, the prize sought by Jason, a prince of Iolcus in Thessaly. Jason's father was usurped by his brother, King Pelias, who promised to cede Jason the throne if he retrieved the Golden Fleece from King Aeëtes of Colchis. The ram had been sent by Hermes to save the children of the king of Boeotia when their stepmother threatened their lives. One child was killed during the escape, but the other flew safely to Colchis on the ram, where he sacrificed the beast in thanks to the gods, and gave its fleece to King Aeëtes,

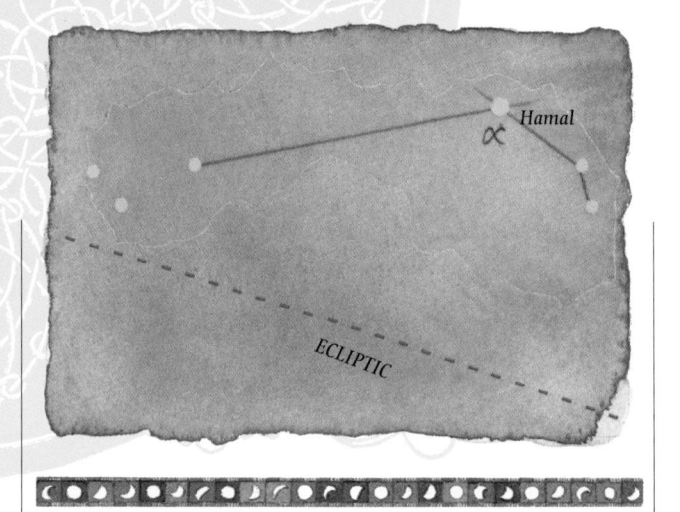

Hamal

α

ECLIPTIC

Aries

who kept it in the grove of the war-god Ares, guarded by an unsleeping dragon.

Jason gathered a crew (the Argonauts) and set sail in the *Argo* to find the Golden Fleece. When the hero reached Colchis, Aeëtes promised to relinquish the fleece if Jason could perform certain difficult tasks. These he achieved, but Aeëtes would not keep his word. However, the king's sorceress daughter, Medea, bewitched the dragon so that Jason could seize the fleece and make his escape back to Thessaly with both the prize and Medea as his bride.

The association of this constellation with the Greek war-god Ares carries through into astrology with its rulership by the planet Mars. The constellation of Aries was also frequently dedicated to Athene and Zeus, and to the "Unknown God".

TAURUS

A star map showing Taurus relative to Orion and Gemini and its location on the ecliptic. The red star Aldebaran (Alpha Tauri) marks the bull's eye and Al Nath (Beta Tauri) the horn.

Taurus, the Bull, is a striking zodiacal constellation which in Northern skies culminates at midnight in early December. Northwest of Orion, it is identified by its two star clusters, the Pleiades and the Hyades, which form a loose group around the first-magnitude star Aldebaran (Alpha Tauri, pale red), the "Eye of the Bull". Close to the ecliptic, Alde-

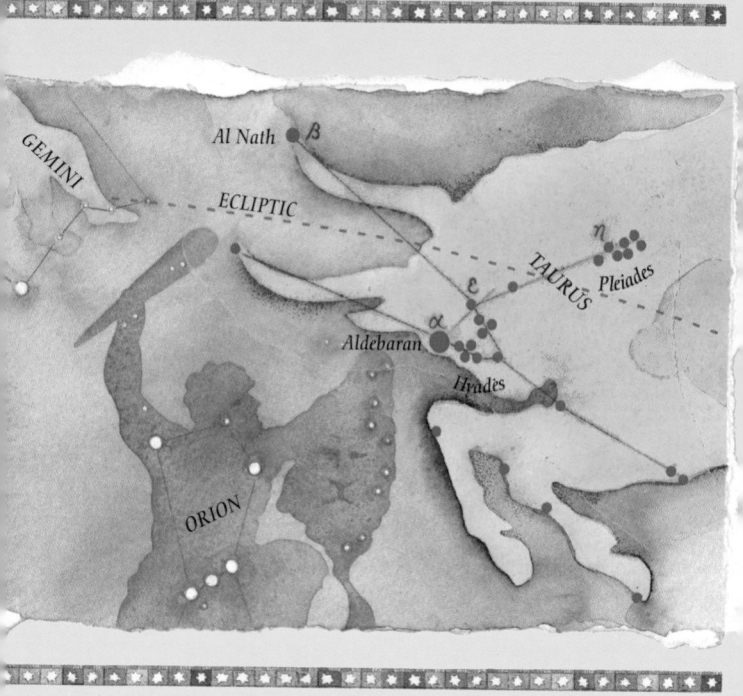

Taurus

baran was one of the four Royal Stars which kept watch on the solsticial and equinoctial junctions on the Sun's path 5,000 years ago, where it marked the spring equinox. The others were Antares in Scorpio (autumn equinox); Regulus in Leo (summer solstice); and Fomalhaut in Aquarius (winter solstice).

The figure of Taurus shows only the front of a bull, his head lowered as if to charge. The northern horn is marked by a brilliant white second-magnitude star, Al Nath (Beta Tauri).

Because it marked the spring equinox from *c.*4000BCE to *c.*1700BCE, when astronomy was founded in Mesopotamia, Taurus was among the earliest recorded constellations. The bull symbolism is a recurrent theme. In Persia the constellation represented the bull slain by the god Mithras. Popular among the Roman legions, Mithraism was a formidable early rival of Christianity.

For the Greeks, Taurus represented the myth of Europa, a beautiful girl abducted by Zeus disguised as a gentle snow-white bull. The girl playfully climbed on its back; it strolled to the seashore but suddenly bounded off into the waves, carrying Europa to Crete, where Zeus ravished her.

Among the Romans, Taurus was sacred to Bacchus, god of wine. During Bacchic

festivals, a bull strewn with flowers was surrounded by dancing girls representing the Hyades and Pleiades. The first of these star groups, the Hyades, formed the 19th lunar mansion Pi in ancient China. Its marker-star was the fourth-magnitude Epsilon Tauri. *Pi* means "hunting net", of the type used to catch birds or rabbits; the warlord Can (Orion) was extended to show him waving this net above his head. In the Mesopotamian

The Pleiades as the Seven Sisters, from an Italian manuscript of the 9th–10th centuries CE. Clockwise from the top, they are: Merope, Celeno, Sterope, Maia, Taygeta, Alcyone. Electra is in the centre.

creation epic, the god Marduk uses the Hyades as a boomerang-like weapon. The Hyades are also associated with the jaw-bone used by Samson to slay a multitude of Philistines. To the Aztecs these stars were an ox's jawbone.

This Roman relief shows the Persian god Mithras slaying the bull. He is ringed by the signs of the zodiac, running anticlockwise.

The Pleiades are located on the bull's shoulder around 15° northwest of Aldebaran. This beautiful cluster of seven stars, like a miniature Plough, occupies an area no larger than the Full Moon. They have fascinated sky-watchers from the earliest times and have often been treated as a star group. The brightest is Alcyone (Eta Tauri, third magnitude, greenish-yellow). Greek tradition widely interpreted the Pleiades as seven sisters, an association found in many other cultures.

URSA MAJOR AND MINOR

For an observer in the Northern Hemisphere, the distinctive shape of the seven stars of the Plough, or Big Dipper, high above Leo, is the most readily identifiable of all constellations. No part of it ever disappears from the night sky at latitudes above 40°N, and most of its stars are circumpolar at latitudes above 30°N.

The Plough is part of the constellation of Ursa Major, the Great Bear, forming the rump and tail of a huge bear, but its importance as a grouping in its own right justifies separate treatment. The pattern suggests a saucepan with a long curved handle; the side of the pan away from the handle is marked by the two brightest (second-magnitude) stars of the constellation, Dubhe and Merak (Alpha and Beta Ursae Majoris). These provide an invaluable reference for the sky-watcher because the arc between them is a virtually precise meridian or north–south line, pointing directly to the

Ursa Major and Minor

Pole Star some 30° away. Their turning through the night serves the navigator as the hour hand of a celestial clock.

There is a fascinating detail in the second star from the end of the pan-handle. On casual observation we see only a single second-magnitude star, Mizar (Zeta Ursae Majoris), but more careful inspection reveals a fifth-magnitude star close by it, to the northeast. This is Alcor (80 Ursae Majoris), and despite its apparent insignificance it has always attracted attention. According to Arab tradition, this star has the lowest rank in the celestial hierarchy (Canopus having the highest). The desert Arabs used the separation of Alcor from Mizar as a test of good vision. In one Teutonic myth, Alcor is the frozen big toe of the giant Orwandil (equivalent to Orion), which Thor broke off and threw into his cart (the Plough).

The identification of the seven stars as the vehicle of a god was predominant

A star map of Ursa Major and Ursa Minor as they are positioned around Polaris (which forms the last star in the tail of the Lesser Bear). The Plough (Big Dipper), part of Ursa Major, is picked out in dark red.

Ursa Major and Minor

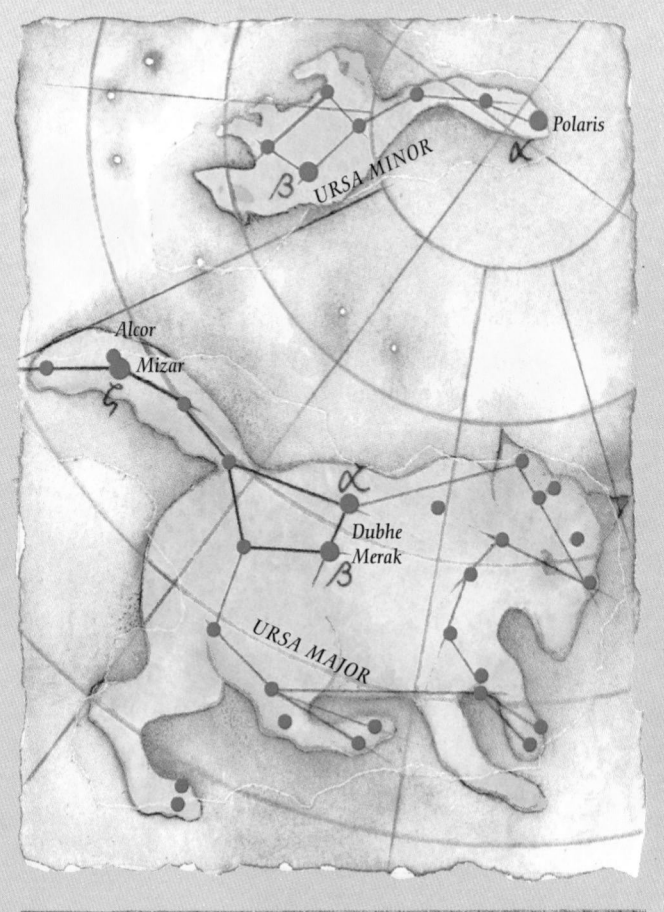

Ursa Major and Minor

in pre-Christian Europe, and is found in Greek and Latin authors, and among the Jews and Arabs. The same image occurs in China, where a relief of *c.*150CE shows one of the Celestial Bureaucrats riding in a carriage of the seven stars.

The Indian treatment of the stars as the Seven Rishis, or sages, has been resonant in modern occultism and theosophy (divine wisdom). The term *rishi* is related to the Sanskrit for "bear", reflecting the myth-complex that has become dominant in our modern view of the wider constellation. It is also found in North America, where some peoples identify the pale star Alcor as a young girl who was transformed into a bear.

The bear imagery of Western tradition dates back to the Greece of Homer and was widely adopted by Classical authors. The Greek myth of the Greater and Lesser Bears tells of the beautiful nymph Callisto, the daughter of King Lycaon of Arcadia. She loved hunting and joined the band of Artemis (the

Roman Diana), the goddess of the hunt, which meant she must stay a virgin. While she rested one day in the forest, Zeus approached her in the guise of Artemis before revealing himself and ravishing her. Callisto became pregnant, and Artemis dismissed her. Callisto gave birth to a boy, whom she named Arcas.

Hera, Zeus's wife, heard about his adultery and cursed Callisto, turning her into a bear. Callisto hid away in the forest while Arcas grew up to be a hunter himself. One day, when he was out hunting, Callisto recognized her son and rushed to embrace him. Startled by the beast, Arcas raised his bow, but Zeus intervened and changed him into a bear also. He swung both mother and son into the heavens, where Callisto became the Greater Bear (Ursa Major) and Arcas the Lesser Bear (Ursa Minor).

Ursa Major, the Great Bear, known in Arabic by its major star name, Dubhe. The illustration comes from an Arabic manuscript that dates from the 18th century.

Ursa Major and Minor

The Milky Way

On a clear moonless night, away from city lights, the hazy band of starlight that we know as the Milky Way can be seen in all its magnificence, stretching across the sky in a vast arc from horizon to horizon. When we look more carefully at this band, especially with binoculars, we can pick out rich clusters of stars in far greater concentrations than anywhere else in the sky. We are looking directly along the plane of rotation of the flat disk of an enormous system of stars. The clouded light is in fact the merged, suffused light of countless stars, too distant to distinguish separately, which make up our galaxy. The Sun is just one of 100,000 million stars in this vast system, around two thirds of the way out from the centre, lying on one of its spiral arms. Outside the band of the Milky Way, virtually all the stars we can see are close members of the

Winding a path across the ecliptic through zodiacal Gemini and Sagittarius, the Milky Way (shown here in a cigarette card) has often been imagined as a great river in the night sky.

same system, but their concentrations are relatively sparse because they fall outside the spiral disk of the galaxy. Our galaxy has a diameter of some 100,000 light years. Beyond it in all directions lie vast wastes of space until we reach other star-systems. At two million light years, the nearest major galaxy to us, visible as a faint smudge about the size of the Full Moon, is the Andromeda galaxy, in the constellation Andromeda. The centre of our galaxy lies in the constellation Sagittarius, where the Milky Way is visibly broader and more dense.

The Milky Way has almost universally been described as a road or river, and in several regions of the world a real river has been seen as its terrestrial counterpart. In ancient Egypt, the Milky Way was identified with the Nile, in India with the Ganges. In China, its name was Sky River and its earthly representatives were the Huang He (Yellow River) and the Han River. Sky River plays its part in the popular legend of the weaver-girl

The Milky Way

The Milky Way

and the herd-boy, represented by the beautiful star-pair Vega and Altair (Alpha Lyrae and Alpha Aquilae). When the girl married a neighbour, who herded cattle and lived on the banks of the Sky River, her father, the Sun-god, grew angry and decided to separate the couple: they were forced to live on either side of the river of stars, which is where we see them today. They are allowed to meet just once a year, on the festival of the seventh day of the seventh month. The Sun-god calls a flock of magpies, who form a bridge across the stream, and the weaver-girl, Vega, runs joyfully across.

A significant theme in various cultures is the Milky Way as the path of souls. The Sumo people of Honduras and Nicaragua believe that Mother Scorpion dwells at the end of the Milky Way, waiting to welcome the souls of the dead. She also gives birth to the newborn and suckles them at her many breasts. The Roman scholar Macrobius

The Milky Way, as seen from latitudes *c*.50°N – the approximate view from London, Toronto and Moscow.

(*c.* 4th century CE) tells us that the souls of the dead ascend through the sign Capricorn and descend through Cancer; this corresponds with the path of the Milky Way where it intersects the constellations Gemini and Sagittarius. He explicitly states that the gateway for the souls is the intersection of the Milky Way and the circle of the zodiac. In Mayan belief, Wakah-ch'an, the "World Tree", was thought to hold up the sky at this same ecliptic-galactic intersection.

The Greek myths offer several different lines of interpretation for the Milky Way. One version of its origin concerns Herakles, the son of Zeus and Alkmene, a mortal woman. Zeus wanted his son to become immortal by the milk of a goddess, so he led his wife Hera to believe that he was an abandoned baby, and induced her to suckle the child. But the lusty infant sucked her nipple with such force that she flung him down in pain. A fountain of milk sprayed from her breast and became the Milky Way.

A Renaissance woodcut showing Hercules (Herakles) suckled by Juno (Hera). The milk that sprayed into the heavens became the Milky Way; the milk that fell to the ground gave rise to lilies (shown here).

The Milky Way

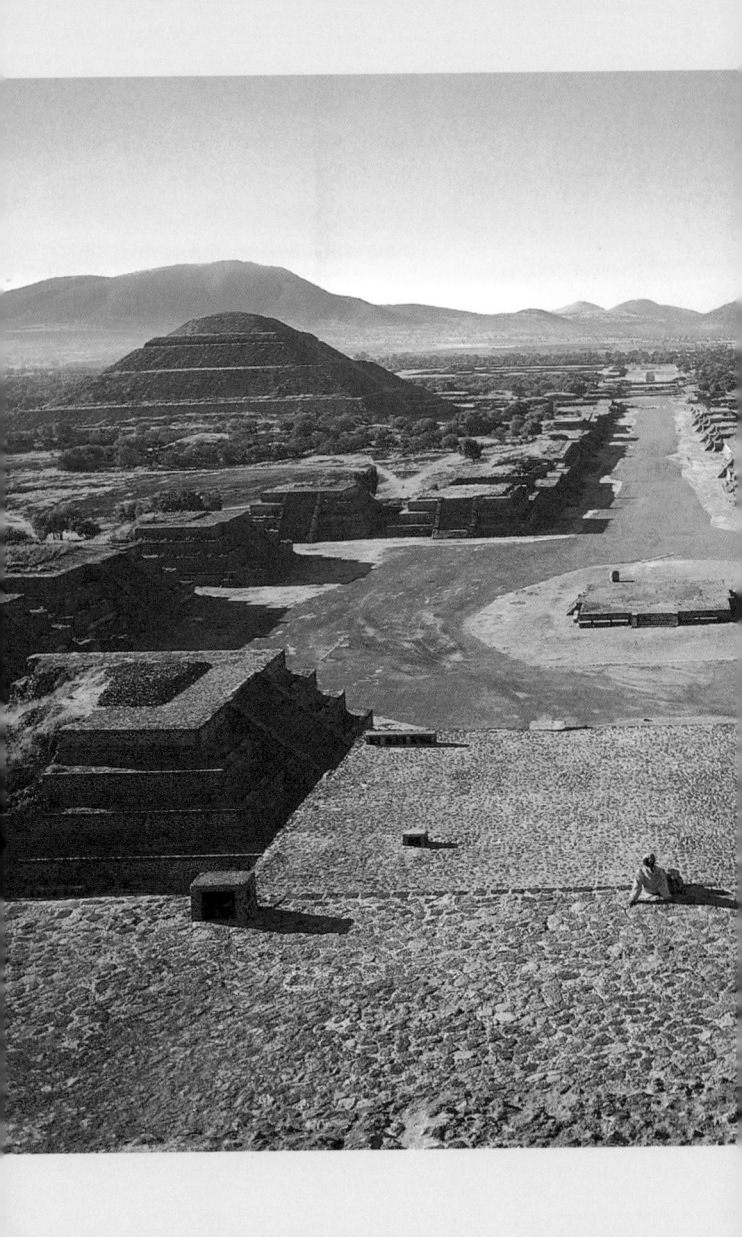

The great urban and ritual centre of Teotihuacán (see pages 323–328), viewed from the Pyramid of the Moon along the "Street of the Dead", with the Pyramid of the Sun on the left. These pyramids align with the sacred peak Cerro Patlachi-que, on the horizon to the south.

SACRED
ALIGNMENTS

Many ancient archaeological sites show compelling evidence of deliberate alignments with heavenly phenomena, especially sunrises at solstices and equinoxes, moonrises and moonsets at major and minor standstills (see pages 30–31), and sometimes with stars or planets. The scientific study of such alignments is called archaeoastronomy. The precise uses of these sightlines must in most cases remain a mystery, but it is reasonable to hypothesize that they related to the continuity of seasonal cycles, often with an underlying emphasis on fertility.

The pyramids of Giza have been shown to relate to Orion and to the Hyades, whose alpha and epsilon stars are shown here.

THE SCIENCE OF ALIGNMENTS

All around the world the ruins of ancient civilizations demonstrate an intimate and intricate knowledge of the science of the skies. Most sites are shrouded in mystery. However, archaeoastronomical detective work may reveal that one site embodies sightlines to midsummer and midwinter sunrises or sunsets while another aligns to the northernmost setting point of Venus (or did so 2,000 years ago).

Once a site had been selected, working out solar and lunar alignments would probably have been a relatively simple, although painstaking, process. A marker at the centre of the monument would have been the starting point. True North could then have been found by marking the setting and rising points of a star on the horizon and finding the midway point between them. The Sun's solstitial rising and setting points and the lunar

The Science of Alignments

The Science of Alignments

The three stand-
ing stones of
Ballochroy, on
the west coast of
Scotland, point
over a burial kist
(a box of stone
slabs) toward
Cara Island,
7 miles (12km)
away, a foresight
to the midwinter
sunset position.
Moreover, a line
from the centre
stone to Corra
Bheinn, a moun-
tain 19 miles
(30km) away
on the island
of Jura, indicates
the position
of the midsum-
mer sunset.

standstill points, it is thought were located by a process of repeatedly marking,
with posts, the rising and setting points
on the horizon, over a period of time.
Permanent markers would then have
been erected to show special points of
interest. Over time, the builders would
probably have gained sufficient understanding of the sky, so would not have
needed to observe, say, a complete 18.6-
year lunar cycle to build a new monument.

Many sites use natural features on the horizon to mark particular points, and these computations would have taken a great deal of time to work out. Very few sites have totally flat horizons, and a hill at a critical point could drastically affect a sunrise or moonset position.

When looking for evidence of ancient astronomical monuments, it is important to remain sceptical: almost any two randomly scattered stones could coincidentally indicate a significant astronomical alignment. There has to be shown more evidence of intent. It is important to establish first the intended viewing point ("backsight"): at Uaxactún, for example, there are distinct viewing platforms, while many megalithic monuments have a clear central focus. From such a fixed point, an alignment can be indicated by, for example, a distant stone or a hollow or hump on the horizon. An outlying feature of this kind is known as a "foresight". The greater the distance between foresight and backsight, the more accurate the alignment can be.

NOTE

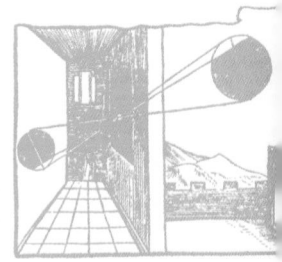

In the following pages, square, coloured symbols are used to indicate the midsummer, midwinter and equinox sunrises and sunsets, and the major and minor standstills of the Moon: the symbols correspond to those used in the explanatory diagrams on page 21 (Sun) and pages 30–31 (Moon).

STONEHENGE
ENGLAND

Stonehenge

An aerial view of Stonehenge showing the present remains of the inner and outer sarsen rings and the two remaining Station Stones. The Heel Stone is off the picture, bottom right.

In 1740 the antiquarian William Stukeley noted that the axis of the great grey stones of Stonehenge on Salisbury Plain in Wiltshire, England, aligned to the northeast "where abouts the Sun rises when the days are longest". However, the belief in a connection between Stonehenge and midsummer goes back much farther, for it was the site of traditional midsummer festivities for many centuries prior to Stukeley's time. Today it is well known that the midsummer Sun rises over the tall outlying Heel Stone when viewed from the centre of Stonehenge.

But the monument's astronomical aspects are more complex than this. Stonehenge has been closely intertwined with the development of archaeoastronomy, and it

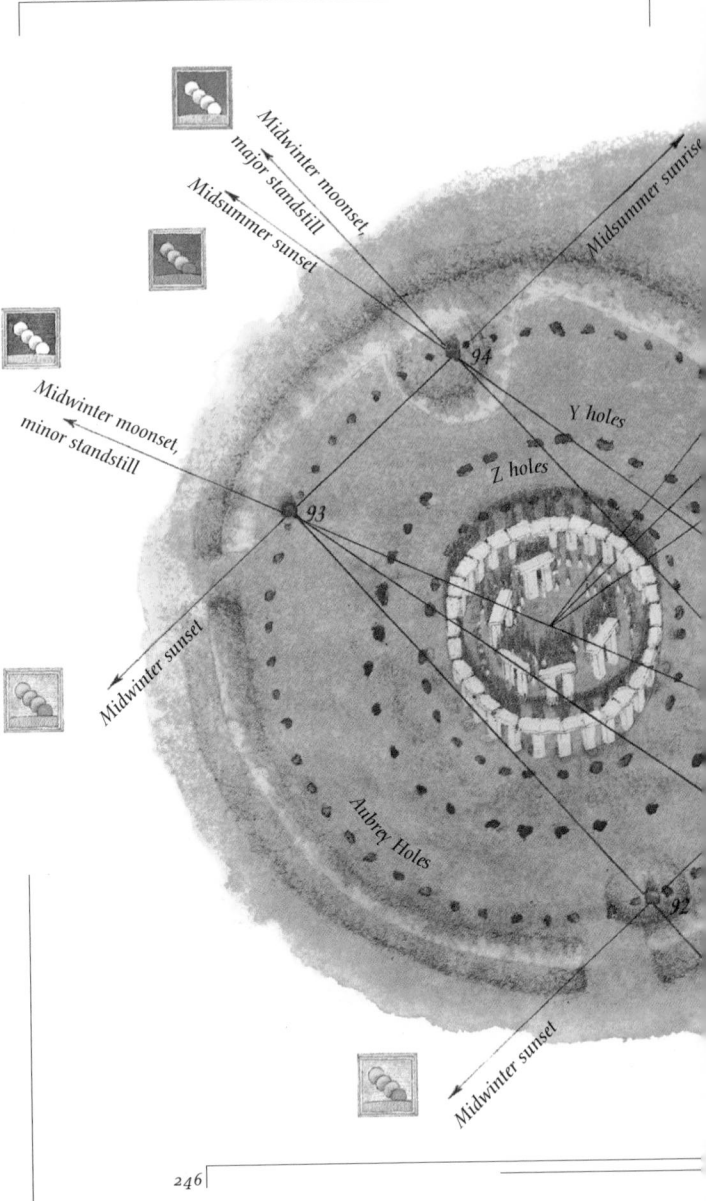

Midwinter moonset,
major standstill

Midsummer sunset

Midwinter moonset,
minor standstill

Midwinter sunset

Midsummer sunrise

Y holes

Z holes

94

93

Aubrey Holes

92

Midwinter sunset

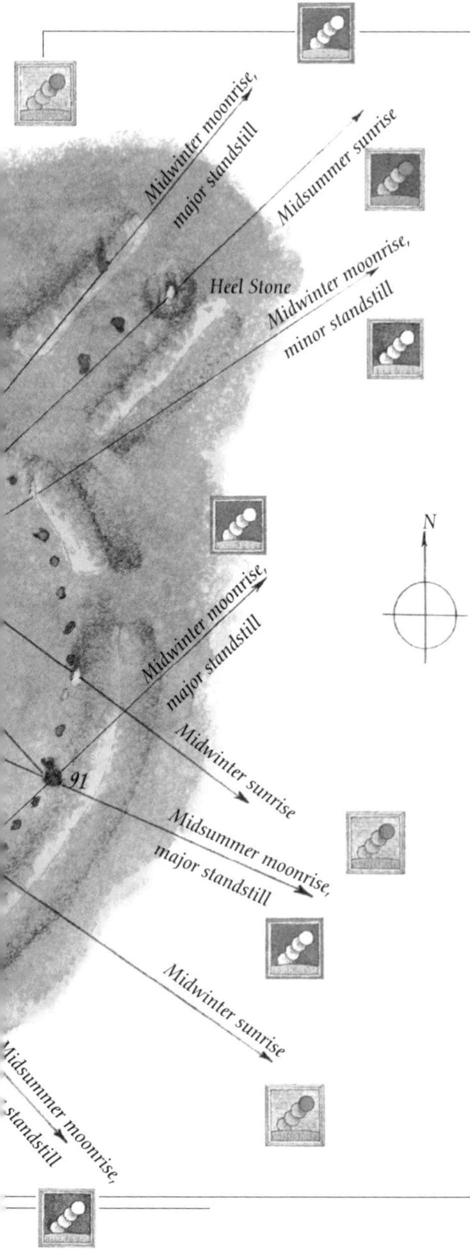

Midwinter moonrise, major standstill

Midsummer sunrise

Heel Stone

Midwinter moonrise, minor standstill

N

Midwinter moonrise, major standstill

Midwinter sunrise

91

Midsummer moonrise, major standstill

Midwinter sunrise

Midsummer moonrise, standstill

Stonehenge

This plan of
Stonehenge
shows celestial
alignments dis-
covered by Lock-
yer, Newham
and Hawkins.
The rectangle
formed by the
Station Stones
(the red dots
marked 91–94)
is traced out by
their alignments.
The outer ring
and inner horse-
shoe of mega-
liths are made
from local sarsen
stone, but how
such massive
stones were
transported and
put into place so
precisely remains
a mystery.

was the so-called "father of archaeoastronomy", the scientist Sir Norman Lockyer, who began the scientific appraisal of this complex site in 1901. Lockyer had previously studied astronomical alignments in Greek and Egyptian temples. He attempted to date Stonehenge by calculating back to the time when the first gleam of the midsummer rising Sun would have been perfectly in line with the monument's axis. This alignment gave Lockyer a range of between 1600BCE and 2000BCE, roughly the date of what is now called Stonehenge III (it was not yet known that there were several versions of the structure: see page 255).

Lockyer saw that the axis could be extended to Silbury Hill in one direction and Castle Ditches in the other, both being hills with prehistoric earthworks. He also calculated that a diagonal across the rectangle formed by the "Station Stones" (see plan, pages 246–247) would give sunrises and sunsets on significant days, such as May 1. Unfortunately, he presented his material poorly and made several errors, and archaeologists disputed his claims. Nevertheless, Lockyer's work was the start of serious archaeoastronomical study at Stonehenge.

More sophisticated studies did not appear until the 1960s, when the careful measurements of C.A. Newham showed that the sides of the Station Stones' rectangle gave alignments to key Sun and Moon rising and setting positions. Newham was not the first to notice some of these, but confirmed their full extent. Another alignment involved the Heel Stone and Station Stone 94, which he claimed would have marked the

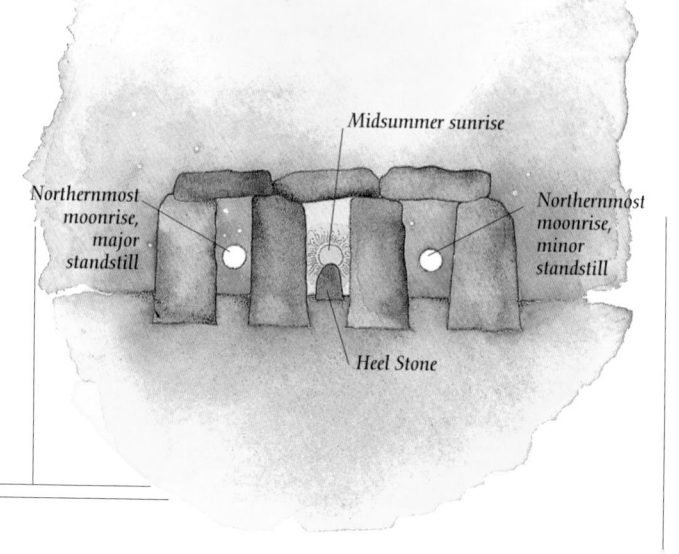

Northernmost moonrise, major standstill

Midsummer sunrise

Northernmost moonrise, minor standstill

Heel Stone

equinox moonrise. Indeed, Newham came to believe that in its earliest stage, Stonehenge "was essentially a site for the investigation of lunar phenomena". He felt that postholes discovered in the entrance causeway were for poles erected by the builders of Stonehenge to mark the passage of the Moon's movements for over a century, so that they could work out the skyline positions of the Moon in the course of its complex cycle of 18.6 years. He also noted that one of the thirty uprights of the outer sarsen circle was thinner than all the others, and suggested that the whole ring might have symbolized the 29.5 days of the lunar month.

It now seems likely that Stonehenge was originally laid out on a lunar axis and that this was deliberately realigned by 4° at a later date to form a solar one. However, if Stonehenge was originally a lunar temple, why does the midsummer Sun rise over the Heel Stone? The fact is that this 37-tonne, 16-foot (5m) tall out-lier does not exactly mark sunrise on the solstice: the Sun rises just to the left (west) of the stone when viewed from the centre of Stonehenge, and this distance would have been greater 5,000 years ago, owing to precession. In 1979, a stone hole was uncovered a few yards to the west of the Heel Stone. If this marked the position of a now-lost stone, then the midsummer sunrise would have been neatly framed by two great stones, of which only the Heel Stone survives. But it is also possible that the hole marks a former position of the Heel Stone itself, which may have been moved when the axis of Stonehenge was changed. It could even be that the midsummer sunrise connection with the Heel Stone is a coincidence. From the centre of Stonehenge the out-lier is viewed through a gap between two uprights in the outer sarsen ring. The two gaps on either side are "windows" toward the northernmost and southernmost points where the Moon rises in its

Stonehenge

Stonehenge

18.6-year cycle (see illustration, page 249). The midpoint of this oscillation happens to be marked by the Heel Stone.

Gerald Hawkins, an astronomer from North America's Smithsonian Institution, became aware of Newham's work and began to investigate the Station Stones' rectangle himself. He found even more alignments associated with the arrangement of the four stones – but to establish this he had to incorporate questionable extra markers, such as holes found during excavations, that may never have held stones or poles. Hawkins also looked at other aspects of the site – in particular, the great sarsen stones that are most often associated with Stonehenge. He found that the inner "horseshoe" of great trilithons (pairs of uprights with a lintel stone across the top) yielded sightlines through the gaps between the uprights in the outer sarsen circle toward important solar and lunar rising and setting positions. The narrow gaps between the

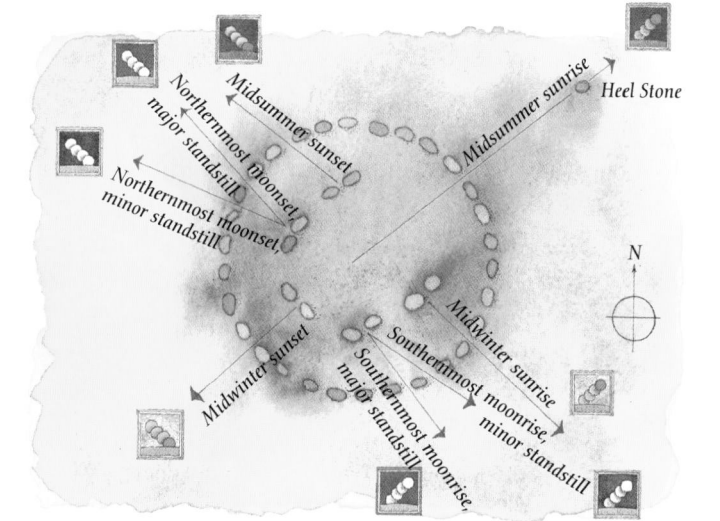

Midsummer sunset
Midsummer sunrise
Heel Stone
Northernmost Moonset
major standstill
Northernmost moonset
minor standstill
N
Midwinter sunrise
Southernmost moonrise
minor standstill
Midwinter sunset
Southernmost moonrise
major standstill

trilithon uprights act like gunsights to the wider gaps in the sarsen ring. None of these lines was extremely precise (the spaces were too wide), but they did provide "windows" that framed astronomically significant sections of the horizon.

Hawkins also noted that the ring of pits known as "Aubrey Holes", just inside the circular ditch around the stones of Stonehenge, numbered 56. The Moon takes 56 years to fulfil its eclipse cycle, and Hawkins proposed that the pits could be the remains of an eclipse

This diagram illustrates Hawkins' theory that the sarsen stones provide sightlines to the most important rising and setting points of the Sun and Moon.

predictor. He envisaged marker stones being moved by one Aubrey Hole a year, with a Moon marker stone moved by one upright a day around Stonehenge's outer sarsen ring, their combinations telling of impending eclipses. The astronomer Fred Hoyle offered another version of this theory using a different system and fewer markers; however, neither idea has convinced archaeologists.

In 1973 the Scottish engineer and astronomer Alexander Thom, whose work over several decades revolutionized archaeoastronomy, finally turned to Stonehenge. With his son Archibald he made a new, accurate survey of the monument and sought longer (hence more accurate) alignments than had hitherto been claimed. Like Lockyer at the turn of the 20th century, the Thoms envisioned Stonehenge as the centre of landscape-wide alignments.

While we can never be sure of the details, there can be no doubt that astronomy played a major part in the structure and uses of Stonehenge. Equally certain is that it was never solely an astronomical observatory. Astronomy would have been used in a cosmological, religious context as an aid to ritual and ceremony – spirit not science, astrology not astronomy. But this fact does not diminish the skill and accuracy of the observations that were made.

THE PHASES OF STONEHENGE

Stonehenge I
The earliest feature on the site, *c.*3300BCE, seems to have been a large round timber building. About a century later a circular ditch-and-bank enclosure (the henge), the Aubrey Holes and, probably, the Heel Stone (and its possible partner) appear. Two stones mark the henge entrance and wooden structures may have been erected.

Stonehenge II
Between *c.*2200BCE and *c.*2000BCE the site is realigned to orient with the midsummer sunrise. Along this line an avenue *c.*500 yards (460m) long approaches the entrance from the NE. An incomplete ring of bluestones is put up within the henge but later taken down, leaving the Q and R holes. The Station Stones may have been set up at this period.

Stonehenge III
IIIA (*c.*2000BCE): Five sarsen trilithons are erected within a lintelled ring of 30 smaller sarsen uprights. Two sarsens at the entrance include the fallen "Slaughter Stone". IIIB (*c.*2000–1550BCE): About 20 bluestones are set up within the trilithons. Holes (Y and Z holes) are dug around the outer sarsens, but seem unused. IIIC (*c.*1500–1100BCE): The bluestones are rearranged and the avenue considerably extended.

AVEBURY
ENGLAND

The Barber Stone, part of Avebury henge, the largest circle of Neolithic standing stones in the world. This stone stands in the southwest quadrant of the outer circle.

Avebury in Wiltshire, England, contains the world's most extraordinary collection of Neolithic (New Stone Age) sites. Although less famous than Stonehenge, some twenty miles (32km) to the south, the Avebury complex is on a much grander scale and is one of the world's best-preserved Neolithic landscapes, its monuments created to blend seamlessly with the natural topography.

The best-known feature of this sacred landscape is Avebury henge, the world's largest stone circle, now encompassing more than half of the village of Avebury. The ditch of the henge, which has an outer bank, was originally 33 feet (10m) deep but has filled up over the ages, although it is still dramatic in appearance. The ditch encloses an area more than 28 acres (11.5ha) in extent.

The surviving standing stones are sited around the inner lip of the henge ditch. Within this huge circle, 1,140 feet (347m) in diameter, are the remains of two smaller stone rings. The centre of the southern circle was marked by the "Obelisk", a huge stone 20 feet (6m) in height and 8 feet (2.5m) in girth. Already fallen when antiquarian William Stukeley sketched it in 1723, it was subsequently broken up and removed. Its original position is marked by a concrete plinth erected in the 1930s.

The remains of a great avenue of stones, the "Kennet Avenue", runs southward from the henge's southern entrance for a mile (1.6km) to a huge palisaded timber structure by the River Kennet, then turns east up Overton Hill to the "Sanctuary". Once, apparently, a group of timber buildings or ritual poles like totem poles, the Sanctuary was finally a ring of stones. Its purpose is unknown, but it may have involved funeral rites: human bones and evidence of feasting

Avebury

Avebury

have been found here. When he saw this stone ring being destroyed, Stukeley was enraged.

The henge at Avebury dates to *c.*2600BCE and is thought to be contemporary with Silbury Hill, a mile to the southwest. Silbury, part of the Avebury complex, is an artificial mound with a flat summit and a substructure of chalk. At 130 feet (40m) high, it is the tallest prehistoric feature of this kind in Europe. Excavations in 1969–70 showed that it was a Neolithic structure and contains no burial or chamber as was first thought. However, archaeologists did find grass almost 5,000 years old preserved – still green – at the heart of the mound, along with flying ants. So while we cannot tell the exact year that Silbury was begun, we can determine that it was in the last week of July or the first week of August!

An aerial view of Avebury showing the distinctive ditch-and-bank feature of the outer rim. The remaining stones of the inner South Circle can be seen clearly, as can those of the outer ring on the south side.

Less than a quarter of a mile (400m) east of Silbury is a natural ridge known as Waden Hill; intriguingly, it is of virtually equal height to Silbury. To the south of Silbury are the West and East Kennet long barrows, among the complex's oldest features. West Kennet has been dated to *c.*3600BCE and was no ordinary tomb. At its eastern end there is a façade of megaliths and an entrance passage into the mound leading to a stone chamber. In side chambers off the passageway, carefully selected human bones were found, perhaps evidence of ancestral rites. At a time when Silbury and the henge would have been in use, the barrow was closed down and the entrance blocked up. However, the chamber complex takes up only a small part of the mound, which stretches for 330 feet (100m) and averages 10 feet (3m) in height. It is thought that it was once shorter. The reasons for lengthening it are a mystery, but at least one theory relates to ancient sightlines (see pages 261–262).

Avebury

Avebury

Overlooking the whole Avebury area is Windmill Hill, a natural feature just over a mile north of the henge. People were already gathering here for unknown purposes *c.*4000BCE, before any stones had been erected.

Through this sacred landscape runs the river Kennet, its source a spring half a mile (800m) from Silbury. The spring probably always had a special status, since it dries up in the winter and starts flowing again, in the direction of the rising Sun, in February. The waters were thought to possess healing properties.

There were no astronomical findings at Avebury until recently. The West Kennet long barrow was oriented to the equinox sunrise, but this may be coincidental, because it is oriented east-west.

The map below shows the Avebury complex, with sightlines (blue arrows) to Silbury Hill from the henge, two of the barrows and the Sanctuary. West Kennet Avenue is shown as two lines of brown dots.

Several researchers have identified other possible astronomical associations for the site, and in 1989 a new theory was proposed by the present writer. It seemed likely that Silbury Hill, rather than the henge, was the heart of the complex. It was like the hub of a wheel, surrounded by the henge, the Sanctuary, the two great long barrows and another at Beckhampton. Viewing Silbury from these various locations, a curious coincidence was noted: the skyline always intersected the profile of Silbury Hill between its flat top and the remains of a ledge, never previously explained, about 17 feet (5m) below the summit.

Viewed from the Obelisk, the very top of Silbury is only just visible between the distant horizon and, in the foreground, the slope of Waden

A plan of Avebury henge showing the positions of existing, fallen and lost stones. The inner North Circle (the Cove) and South Circle are outlined with dotted lines. Pale areas are roads and buildings.

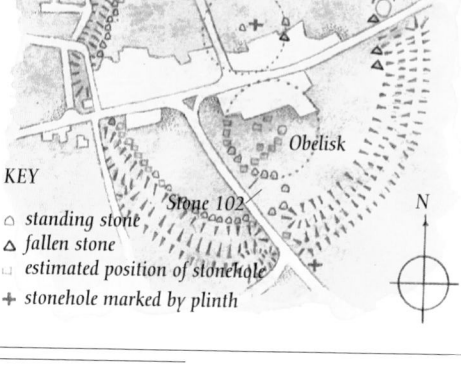

The Cove

Obelisk

KEY

Stone 102

⌒ standing stone
△ fallen stone
⊔ estimated position of stonehole
✛ stonehole marked by plinth

N

Hill. This sightline may have depended on the harvest, because when the cereal crop on Waden Hill is at its height, it virtually conceals Silbury. The cereals grown at Neolithic Avebury are known to have been taller, and the view would have been blocked completely when it was time for harvest. This might relate to the fact that Silbury was begun in late July/early August, traditional harvest time (Lammas in the Christian calendar, Lughnasa in the Celtic calendar). As Michael Dames has argued, Silbury may have been a "harvest hill" sited close to the Kennet's source to symbolize the bountiful Earth Mother.

The skyline coincidence occurs from West Kennet long barrow, but only from the western tip. Archaeologists suspect that the extension to the already ancient barrow probably took place at the time Silbury and the henge were built. Is this why the mound was extended? Also, the skyline that intersects the profile of Silbury from this point is actually formed

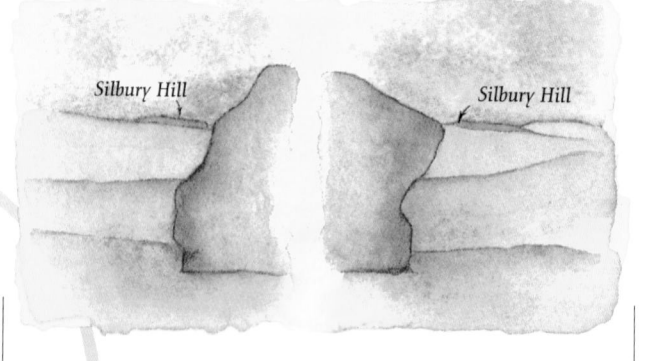

Silbury Hill   Silbury Hill

by the bulk of Windmill Hill. This sightline therefore links all the various periods that the complex embraces.

What was the reason for this skyline coincidence? What was the significance of Silbury's ledge? Looking east from the flat summit of Silbury, Waden Hill looms large in the foreground, with the distant horizon of the Marlborough Downs visible slightly above it. The present writer saw that Silbury was just high enough to separate these near and far skylines, and that the contours of Waden Hill remarkably mimicked those of the far horizon, except for one segment, where the distant skyline dipped slightly. Seen from the ledge beneath Silbury's summit, the

Two views of Silbury Hill looking southward from the Obelisk stone (from the east side in the left picture, the west side in the right) in Avebury henge. The stone in the foreground is Stone 102 (see plan, page 261). Waden Hill conceals most of Silbury from this position, especially at harvest times when crops are high.

Avebury

far horizon exactly coincided with the top of Waden, but the dipped section dropped just behind it. It was calculated that this section gave a "window" for the rising Sun at Lammas/Lughnasa (and around May 1, the Beltane of the Celtic calendar, when the Sun also rises on that part of the eastern horizon): a "double sunrise" effect was likely at these times if the dawn was viewed first from Silbury's summit and then from its ledge.

Observations on August 1, 1989, revealed just how dramatic this effect actually was. From the top of Silbury the Sun rose over the distant horizon. When the observer moved swiftly down to the ledge, the Sun seemed to rise for a second time, a couple of minutes later, over Waden Hill: a ceremonial sunrise effect.

There was a further surprise in store. Looking west from Silbury's summit, the observer saw a long ray of golden light emerging from the shadow of the mound, as if Silbury, symbol of the Earth Mother, were blessing the crops and the land. This effect is an optical illusion caused by refraction in dew. If you stand in a field with your back to the sunrise, you will see a glow around the shadow of your head, known as a "glory". The "Silbury Glory" is a magnified version of this effect and can only happen through the sunrise "windows" of Lammas/Lughnasa and Beltane.

The evidence, then, suggests that Silbury represented the Earth Mother, cosmologically uniting Earth and sky within the sacred landscape of Avebury.

An illustration of the "double sunrise" effect at Silbury at harvesttide (Lammas/ Lughnasa) and around May 1 (Beltane). Seen from the summit of the hill (here shown schematically), the Sun rises over the far horizon. If one moves down to the ledge, it rises again over the brow of Waden Hill a few minutes later.

Avebury

1. Sun rises over distant horizon

2. Sun rises over Waden Hill

←"Sunrise"→
Window"

CASTLERIGG
ENGLAND

The Lake District of northwest England is a landscape of jewel-like lakes and rounded peaks. Near Keswick stands Castlerigg, a stone circle at least 4,000 years old. It is 110 feet (33.5m) across its longest diameter (it is a slightly flattened circle) and all but five of its 38 stones still stand. Ten stones form an enclosure (the Cove) within the main circle.

Alexander Thom's study of Castlerigg revealed seven solar and lunar alignments. He was convinced that the flattened circle plan, which he had identified at other sites, was deliberate. Most striking was the builders' skill in incorporating the sightlines into this groundplan, because the

Latrigg

Castlerigg

mountainous landscape has highly varied horizon heights, so that the Sun or Moon can rise or set at all sorts of odd times.

In 1976, the artist and photographer John Glover set up his camera within the circle to photograph the midsummer sunset over Latrigg ridge. To his amazement, he saw that the setting Sun caused the tallest stone to throw a long, regular "shadow path". It happens that the gradient from the ridge to the base of the stone continues in the sloping ground beyond the site. This slope enhances the shadow's length, which could thus reach up to 2 miles (3km) beyond the circle. Whether the builders stumbled across a place where they could mesh the variable horizon, astronomy and groundplan geometry, and find a slope at the right gradient for the midsummer shadow, or whether they engineered the slope down which the shadow runs, remains a mystery.

The Castlerigg stone circle during a winter sunset. The stones are local, with the tallest almost 7 feet (2m) high and weighing about 15 tons.

The effect of the midsummer setting Sun over Latrigg, throwing the long shadow of the tallest stone down the hillside. The gradient and scale are here greatly exaggerated for clarity.

Shadow path

MAES HOWE
SCOTLAND

Wild and windswept, Maes Howe is a prehistoric chambered mound on Mainland, the largest of the Orkney Islands off northern Scotland. It incorporates a passage opening into a high, impressive chamber with remarkably intact and finished stonework. There are small recesses off the main chamber. No finds have been made here, but the mound has been raided numerous times. Vikings, for example, left runic graffiti and a beautiful engraving of a dragon.

The passageway of Maes Howe is aligned to the midwinter setting Sun, when its dying, golden rays flood the chamber. In 1894, Magnus Spence, a schoolmaster on Orkney, noted the midwinter sunset orientation of Maes Howe, but further realized that if the line of the passage's orientation was extended it went through Barnhouse, a tall standing stone visible from the entrance to Maes

Howe. Spence also discovered that Barnhouse, a monolith called the Watchstone and the centre of the huge Ring of Brogar stone circle formed an alignment to the midwinter sunrise.

In Spence's time, fires were still lit on Orkney hilltops to celebrate ancient Celtic festivals. He found that some of these hills marked rising and setting positions of the Sun along the alignments linking the stones of Orkney. To him, the ancient Celtic solar calendar was encoded in the very landscape.

The Maes Howe mound in its wild landscape setting. The passageway of this tomb aligns to the midwinter sunset – and to a sightline that passes through the tall standing stone named Barnhouse.

Maes Howe

CALLANISH
SCOTLAND

Callanish, on the northwest coast of Lewis in the Outer Hebrides, is an evocative prehistoric structure that distils all the poetry of the Moon and its cycles. The main site, Callanish I, has some fifty standing stones laid out in a design resembling the later Celtic cross. The central circle of 13 stones measures 41 feet (12.5m) at its widest and within it are the remains of a chambered cairn. Running north is a dramatic avenue of stones. To the south, a row of stones runs in a north-south direction. Shorter stone alignments run asymmetrically to the east and west of the circle.

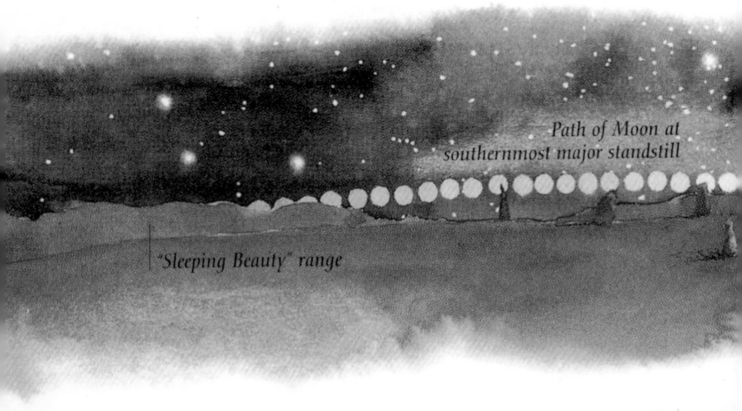

Callanish

Path of Moon at southernmost major standstill

"Sleeping Beauty" range

Archaeoastronomers Gerald and Margaret Ponting made the remarkable discovery that at the time of its southernmost major standstill (see pages 30–31), the Moon skims the stones of Callanish as it travels on a low arc across the horizon after rising in the Pairc Hills (sometimes called the "Sleeping Beauty", or in Gaelic, Cailleach na Mointeach, "Old Woman of the Moors"). The Moon sets on nearby Harris: seen from the end of the Callanish avenue, it appears to sink into the stones of the central circle.

The Pontings also observed in this landscape what is called "re-gleaming", when the Moon sets into the side of a hill then briefly flashes into view again as it passes a "notch" in the hill.

The path of the Moon at its southernmost major standstill, seen from the end of the Callanish stone avenue. Every 18.6 years the Moon is symbolically born out of the Earth Mother, represented by the "Sleeping Beauty" hills, and dies into the sacred stone circle. As the Moon sets behind the stones, there is a "re-gleaming" effect, where it reappears briefly on the horizon.

Callanish

LOUGHCREW
IRELAND

Loughcrew

Two of the cairns that make up the Loughcrew complex near Oldcastle, Co. Meath, Ireland, receive mysterious visitations from the Sun – echoing the legend that Loughcrew was the cemetery of Tailtiu, foster mother of the Sun-god Lugh. The cairns are primarily Neolithic chambered mounds made up of heaps of rocks. They are dotted for approximately 3 miles (5km) along the spine of the Loughcrew Hills.

The most important of the two cairns is Cairn T, on Carnbane East. It is 120 feet (36.5m) in diameter, and was once faced in white quartz. From the entrance, facing just south of east, a passage leads to a main chamber with three

Rectangles of light pass over Stone 14 during the days around the equinoxes, until the sunbeam perfectly frames the main solar symbol.

side chambers. Both passage and main chamber have richly carved stones, notably the backstone (Stone 14) of the end recess in the main chamber. Its prehistoric designs include Sun-like motifs.

The North American researcher Martin Brennan discovered in 1980 that on the days around the equinoxes, rays of the rising Sun probe into the passage until they reach Stone 13 (which has a small "sunburst" carving and juts out into the passageway), when they suddenly burst into and across the main chamber, reaching Stone 14. The configuration of stones in the passage plus a sill-stone (a horizontal stone fixed edgewise into the ground) a short distance in front of Stone 14, shape the beam of light into a fairly regular rectangular form. As the Sun rises, this rectangle passes slowly down the face of Stone 14, ultimately framing perfectly the most complete of the stone's solar symbols. Its eight rays may represent the eight primary markers of the solar year: the solstices, the

Loughcrew

equinoxes, and the "cross-quarter days" (in early February, May, August and November – Imbolc, Beltane, Lughnasa and Samhain in the old Celtic calendar). Brennan noted that the shape of the sunlit rectangle differs between the spring and autumn equinoxes, because the angle of the sunbeams is higher in spring and lower in autumn.

From 1986 to 1990 Tim O'Brien conducted further observations and photographed the light as it passed across Stone 14. Around the main symbol, he noted incised vertical lines with regularly-spaced horizontal lines across them, like grading or scale marks. He discovered that the horizontal lines marked the edges of the rectangular patch of sunlight as it shifted position from day to day down the stone. He confirmed that beams could reach the stone only on the days around the equinoxes, and showed that the builders could have measured solar motion with sufficient accuracy to identify the leap-year cycle of four years.

NEWGRANGE
IRELAND

Newgrange, a gigantic chambered mound in County Meath, Ireland, dates to at least 3200BCE and is one of the world's oldest roofed structures. Once known as Bru na Bóinne, the palace by the Boyne, it was where the ancient Lords of Light were said to dwell, an association reflected in the fact that the monument plays host to the rising midwinter Sun.

The mound is 36 feet (11m) high and 300 feet (90m) in diameter, and around its base are 97 kerbstones, three of which are decorated with carvings. Kerbstone 1,

A plan of the mound, showing the location of the passage entrance into the chamber and the light-beam of the winter solstice sunrise.

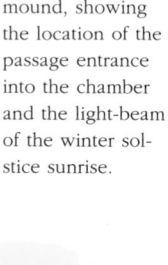

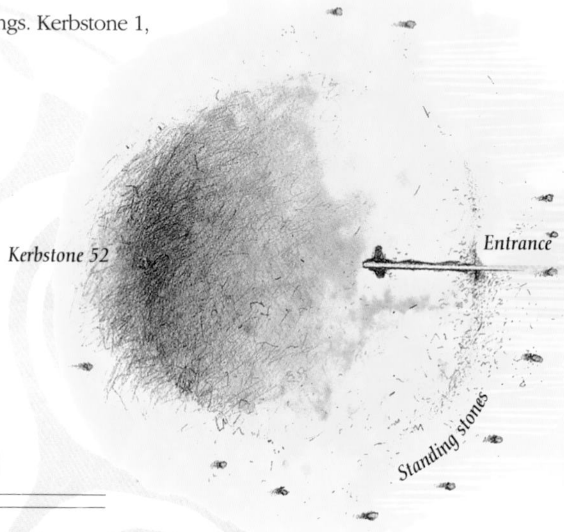

Kerbstone 52

Entrance

Standing stones

richly incised with spiral and lozenge designs, is on the southeastern side of the mound, at the entrance to a passage which extends 62 feet (19m) into the monument to a corbelled stone chamber, 20 feet (6m) high. The passage is about four feet

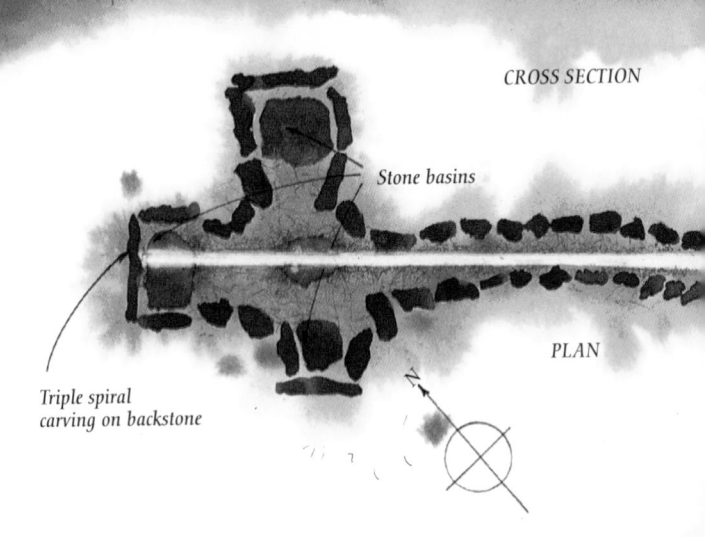

CROSS SECTION

Stone basins

PLAN

Triple spiral carving on backstone

(1.5m) in height, and its floor rises steadily from the entrance to the chamber, which has three side-chambers or recesses. There are stones decorated with rock art in the passage and in the side recesses, including a famous triple spiral carving in the end recess. Surrounding the entire structure is a large, but incomplete, stone circle.

In 1909, Sir Norman Lockyer pointed out that Newgrange (then a ruin) was oriented to the midwinter solstice. Exca-

A cross-section and plan of Newgrange, showing the beam of the rising midwinter sun piercing the roof-box, passing down the passage and onto the backstone.

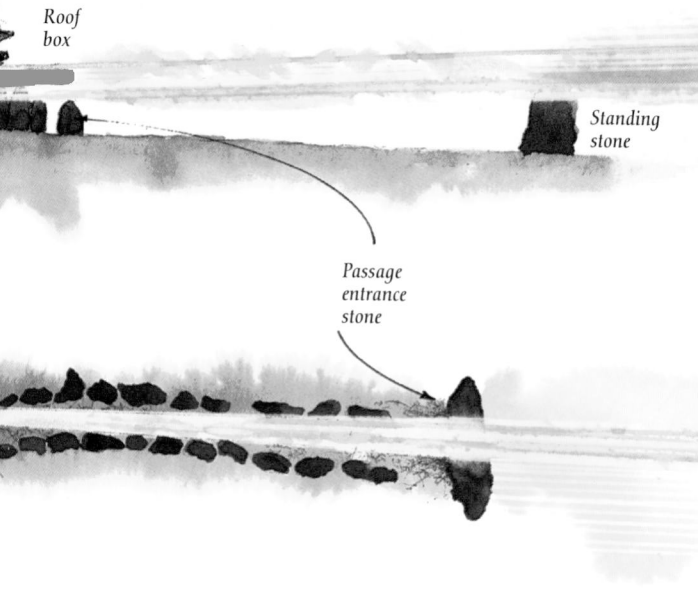

Roof box

Standing stone

Passage entrance stone

Newgrange

vations in 1962–75 under Michael J. O'Kelly uncovered a curious rectangular opening above the passage entrance. O'Kelly suspected that this "roof-box" was designed to admit a shaft of sunlight

into the chamber: a line drawn from the back of the end recess in the central chamber, through the roof-box, aligns to the midwinter sunrise 5,000 years ago. Direct sunlight cannot reach the chamber through the passage entrance because of the upward slope of the floor and the arrangement of the stones lining the passage.

On December 21, 1969, O'Kelly became the first modern observer of the Newgrange sunbeam phenomenon. A pencil-beam of sunlight shot through the roof-box, along the passage and across the chamber floor as far as the basin stone in the end recess. Over the millennia, the Sun's rising position at

The entrance stone and roof-box. The carved patterns are easy to see, although a vertical line dividing the stone in half is now very faint.

midwinter has moved slightly, so that the sunbeam cannot now reach the very back of the end recess. But the effect is scarcely diminished. For 17 minutes, O'Kelly observed,  "the tomb was dramatically illuminated."

Further studies by Tim O'Brien in 1986–88 included a sequence of photos that show the sunbeam sweeping across the floor of the chamber like the glowing hand of a clock. He noted that two of the passage stones lean toward one another to form a triangle that sculpts the sunbeam from the roofbox. Finally, in 1989, astronomer Tom Ray provided statistical proof that the midwinter event at Newgrange could not possibly occur by chance.

A detail of the spiral (possibly solar) markings on Kerbstone 52.

Newgrange

GAVRINIS
FRANCE

The chambered mound of Gavrinis is situated on a small island off southern Brittany in the Gulf of Morbihan. In Neolithic
times, before being inundated by the
sea, this gulf was a fertile plain
with wide rivers and a considerable population.

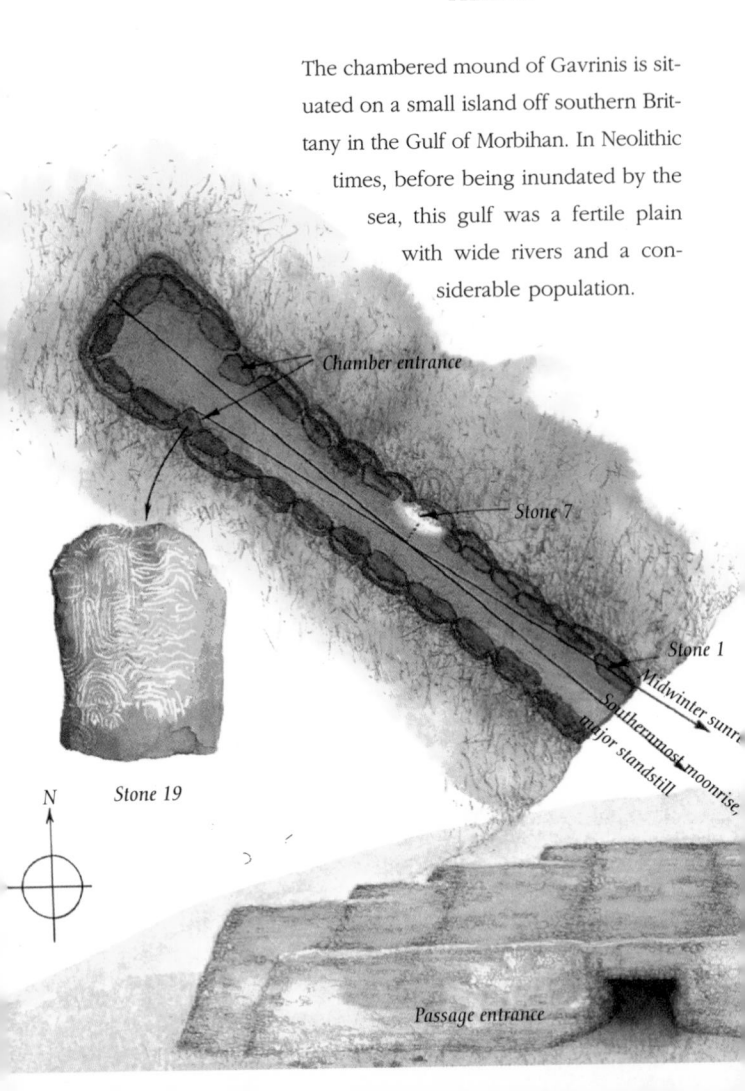

Chamber entrance

Stone 7

Stone 1

Midwinter sunrise

Southernmost moonrise,
major standstill

Stone 19

N

Passage entrance

Dated to *c.*3500BCE, the Gavrinis mound is a rocky cairn over 26 feet (8m) high. Its 6½-foot (2m) high entrance passage runs for almost 39 feet (12m) before opening into a rectangular chamber. Most striking, though, are the densely packed, linear patterns that cover 23 of the 29 upright, flat stones forming the walls of the passage and chamber, making this the most richly carved chambered mound in Europe. The motifs include concentric rings and curves, chevron patterns, serpentine wiggles and triangular shapes.

Some experts have seen stylized goddesses, axeheads and ferns in these flowing lines, but more than anything, the overall effect is like the "psychaedelic" art of the 1960s. Some archaeologists consider the markings to be depictions of optical effects produced by the brain during hallucinogenic trance. Hitherto unexplained Neolithic pottery items are now being interpreted as braziers for burning cannabis. Such mental patterns

OPPOSITE A plan of the chamber at Gavrinis, showing the solar and lunar alignments, the quartz Stone 7 and the carving on Stone 19 at the passage entrance.

BELOW A frontal cross-section of the Gavrinis mound. The layered stone structure is entirely underground: only the passage entrance is visible from outside.

Gavrinis

might have seemed charged with magic and worthy of being set in stone.

A view up the passage from the left side of the chamber entrance toward Stone 1 at the passage entrance gives an orientation to the midwinter sunrise. But the main axis of the passage is aligned toward the southern-most moonrise at major standstill. These solar and lunar lines intersect halfway along the passage, level with Stone 7 – one of the few undecorated stones. This stone is rock crystal and may have glowed in the rays of the rising Sun and Moon at the times in question. The impact of this effect would have been signifi-cant; especially if those who witnessed it were in trance states.

The astronomy is therefore only one component of a whole spiritual and ritual environment.

Gavrinis

Er Grah
FRANCE

Er Grah, also known as Le Grand Menhir Brisé ("The Great Broken Menhir") or Pierre de la Fée ("Fairy's Stone"), lies on a peninsula in the Bay of Quiberon, Brittany. Once more than 66 feet (20m) long, this gigantic stone now lies in four pieces with its tapering top toward the east; its roughly oval cross-section is the result of being smoothed into shape. The collective weight of these blocks has been estimated at up to 342 tons. The stone is not local, but a harder granite, the nearest known source of which is almost 2.5 miles (4km) distant. No one knows when the megalith fell. Some have claimed that it was brought down by lightning or an earthquake, or that it fell and shattered while being erected.

If Er Grah had ever stood, it would have been the tallest standing stone in existence. Professor Alexander Thom felt sure that the stone had once served

Er Grah

as an astronomical foresight (see p.244), visible from points all around the bay, for observing all eight extreme rising and setting points in the minor and major standstills of the lunar cycle. With his team, Thom surveyed the eight directions along which unobstructed sightlines would have had to exist, and confirmed that the stone would have been visible from various points along all those sightlines. As for the observation points (backsights), the team identified possible locations on just five of the lines, although at least one of these has since been shown to be inaccurately placed.

Critics have pointed out that there are so many megalithic remains in the area that some were bound to fall on any lines drawn from the great stone. Oth-

ers have said that Thom's "backsights" are all too dissimilar in type, ranging from passage-graves to single stones, to be purpose-built as viewing stations. But valid or not, Thom's hypothesis remains the only attempt so far to explain the enigma of this ancient megalithic giant.

The area around Er Grah and the alleged back-sights found by Thom and his team. The stone would have been visible from great distances.

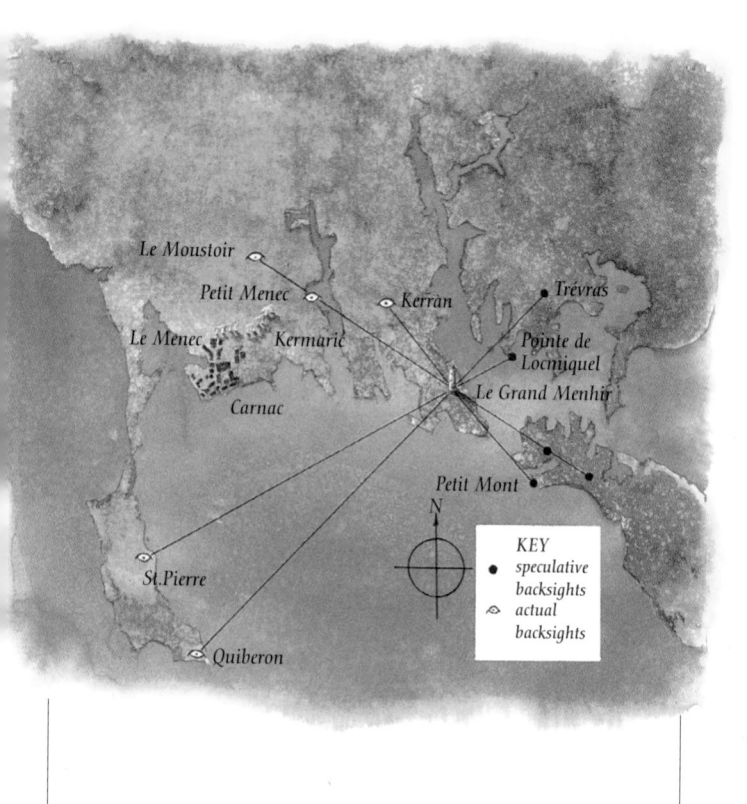

Le Moustoir

Petit Menec   Kerran   Trévras

Le Menec   Kermarié   Pointe de
                                      Locmiquel
        Carnac   Le Grand Menhir

                        Petit Mont

                N

St.Pierre   KEY
                                ● speculative
                                  backsights
                                ◉ actual
    Quiberon   backsights

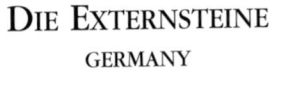

DIE EXTERNSTEINE
GERMANY

A group of five weathered sandstone fingers, each about 100 feet (30m) high, the Externsteine, in the Teutoberger Wald district near Detmold in North Rhine-Westphalia, Germany, are weirdly shaped, suggesting some ogre-like creature of the imagination.

It is claimed by some researchers that the site was a centre for pagan worship until Christianization in 772CE by the emperor Charlemagne, who it is said destroyed an image of the pagan "Tree of Life"

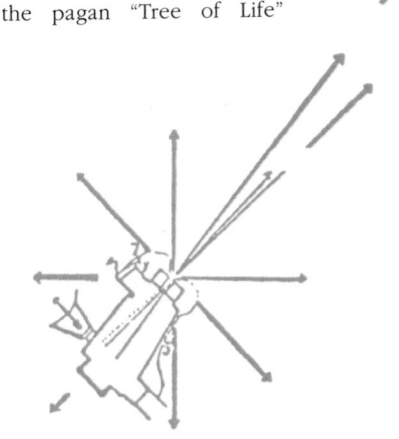

Moonrise, major standstill

Midsummer sunrise

A 1920s map by Wilhelm Teudt showing the orientation of the chapel on Tower Rock to the rising Sun at midsummer and the major standstill of the Moon.

Die Externsteine

adorning the summit of one of the rocks. Others have suggested that Christian worship took place here from the early period of Christianity. Whatever the truth, the site presents a bizarre appearance, with steps, apparently going nowhere, carved in the sandstone pillars; a feature like a sarcophagus hewn out of one giant boulder; and caves or "rooms" honeycombing the lower reaches of some of the stones.

Fairly recent pagan usage is evidenced by an overhanging rock segment that from certain angles looks like a man with his arms raised, as if tied to the rock. It has been proposed that this is the Germanic god Odin hanging on the World Tree. While critics have rightly pointed out that this is a natural feature, it is known to have been worked long ago with tools, presumably to enhance its human appearance.

The site must have been Christianized at some point, because another rock has a relief of Christ's descent from

Die Externsteine

the cross, in which one figure stands on a bent Tree of Life, as if to show the victory of Christianity over paganism.

The most fascinating feature is a chapel hewn out of the rock high on the central pillar known as "Tower Rock" This may have been a pagan chapel partially destroyed during Christianization, or it may have been an ancient Christian chapel. An alcove on its northeast side contains a small "altar" with a circular window behind. It has been noted from at least 1823 that the rising midsummer Sun shines through this aperture when viewed from a niche in the opposite wall. Nowadays the effect is lost because the walls and roof are missing, but when the chapel was enclosed, this would have been a dramatic sight from the dark interior. On the altar there is a slot, into which a crystal or a gnomon was perhaps fitted, to throw rainbow light-beams or a shadow into the niche. The window was also believed to frame the Moon at its northernmost rising position.

THE GREAT PYRAMID
EGYPT

The Great Pyramid was the tomb of Khufu (Cheops), who, like all Egyptian pharaohs, was believed to be divine. His mummified corpse lay in the "King's Chamber"

One of the Seven Wonders of the ancient world, the Great Pyramid stands with other pyramids on the Giza plateau southwest of Cairo. Today's vast edifice would have been even more commanding in its pristine state, faced with smooth, gleaming white stone and topped by a gilded tip to catch the Sun's rays. The pyramid has captured imaginations throughout history, spawning countless theories, fantasies and fallacies. After one has cut through all these there are very few hard data concerning its astronomical function. Like the nearby Sphinx, it guards its secrets well.

The pyramid was built *c*.2600BCE as the

The Great Pyramid

To Orion's Belt 44° South shaft

North shaft

To upper culminati[on] of Thuban 31°

King's Chamber

tomb of the Fourth-Dynasty pharaoh Khufu (called Cheops by the Greeks). The legendary precision of its construction is fully justified: the north and south sides are aligned east-west to an accuracy of less than 2.5 arc minutes, and although the pyramid's base covers 13 acres (5ha), the discrepancy in the lengths of any of the sides nowhere exceeds 8 inches (20cm). That the four

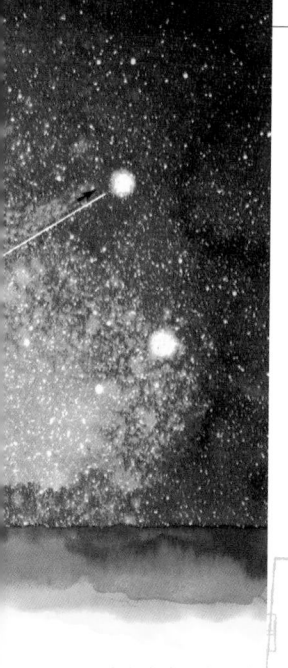

A cross-section showing the stellar alignments of shafts in the Great Pyramid. The south shaft aligns with Zeta Orionis in Orion's belt; the north shaft aligns with Thuban – the Pole Star to the Egyptians.

sides of the pyramid are set so accurately to the cardinal directions indicates that the ancient Egyptians were able to make precise astronomical measurements and transfer them with great skill to the ground. The King's Chamber, deep within the bulk of the pyramid, is also oriented to the cardinal directions.

The internal passages and shafts of the Great Pyramid have also come under scrutiny. The Egyptologist-astronomer team of Professor Alexander Badawy and Dr Virginia Trimble suggested in 1964 that the so-called "ventilation shafts" connected to the King's Chamber may have considerable astronomical and cosmological significance. There are two of these shafts, opening respectively in the north and south walls of the King's Chamber, and leading all the way through the mass of the pyramid to the outside. No other Egyptian tombs are known to be ventilated. The team found that the north shaft aligned to the star Thuban at its culmination (highest point

The Great Pyramid

The Great Pyramid

in the sky), while the south shaft aligned to the transit across the sky of the three stars forming Orion's belt *c.*2700–2600BCE. According to the Pyramid Texts – funerary writings that were inscribed inside pyramids of the Fifth and Sixth Dynasties (*c.*2500–2170BCE) – Thuban was the equivalent of the Pole Star: its description as an "Imperishable Star" in ancient Egyptian terminology means that it was circumpolar. It was to the Imperishable Stars that the pharaoh journeyed after death, there to "regulate the night" and "send the hours on their way". The pharaoh's

HYADES

Epsilon Tauri

The Bent Pyramid

The Red Pyramid

The Giza pyramids have been claimed to provide a "map" of Orion and the Hyades. The Nile mirrors the Milky Way remarkably.

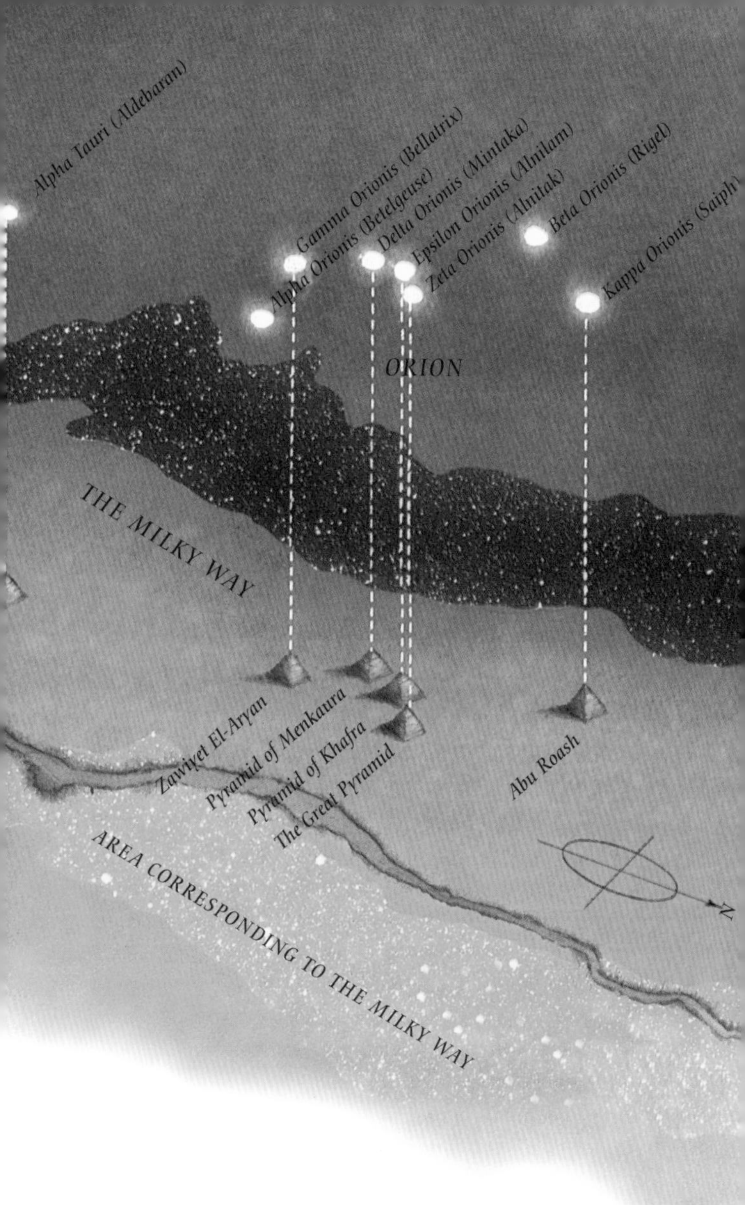

Alpha Tauri (Aldebaran)

Gamma Orionis (Bellatrix)

Alpha Orionis (Betelgeuse)

Delta Orionis (Mintaka)

Epsilon Orionis (Alnilam)

Zeta Orionis (Alnitak)

Beta Orionis (Rigel)

Kappa Orionis (Saiph)

ORION

THE MILKY WAY

Zawiyet El-Aryan

Pyramid of Menkaura

Pyramid of Khafra

The Great Pyramid

Abu Roash

N

AREA CORRESPONDING TO THE MILKY WAY

spirit also ascended to Orion, which to the Egyptians symbolized Osiris and the cycle of birth, life, death and resurrection. One with Osiris, the dead pharaoh maintained the round of the seasons.

Badawy and Trimble therefore raised the possibility that the shafts leading from the King's Chamber were to allow the pharaoh's soul to travel to the heavens. A new twist has been given to this idea by the Egyptologist Robert Bauval, who discovered that five of the seven Fourth-Dynasty pyramids on the Giza plateau – the Great Pyramid, Abu Roash, the pyramid of Khafra (Chephren), the pyramid of Menkaura (Mycerinus) and Zawiyet el-Aryan – were apparently built relative to one another in the pattern of the stars of Orion. Their configuration symbolically brought the god to Earth. Bauval calculated too that the south shaft of the Great Pyramid was aligned with Zeta Orionis in Orion's belt, and that the Great Pyramid represented that star in the "map" on the ground.

KARNAK
EGYPT

Karnak is a large temple complex based around the temple of Amun, on the northern fringe of modern-day Luxor, ancient Thebes. Before the Middle Kingdom (*c.*1980–1630BCE), Thebes was not especially significant, but it did host a local cult of the god Amun. When the throne of Egypt passed to Theban kings in the Middle Kingdom, the city grew in importance and the temple of Amun was enlarged and embellished, becoming Egypt's leading religious centre.

The Karnak complex is composed of the Amun temple plus smaller temples and a range of shrines and chapels, all surrounded by a mudbrick wall enclosing an area half a mile square (1.3 km²), called the Precinct of Amun. Only a few structures from the Middle Kingdom now remain, and most of what can be seen dates to the New Kingdom (*c.*1539–1075BCE). Entrance is from the

Karnak

west, the direction of the nearby Nile, along a causeway flanked by ram-headed sphinxes.

Visually, the dominant feature of the temple of Amun is its main axis, on which stand the various pylons (ceremonial gateways), halls and shrines added by successive pharaohs, and the sanctuary, the oldest and most sacred part of the temple. Sir Norman Lockyer

Karnak

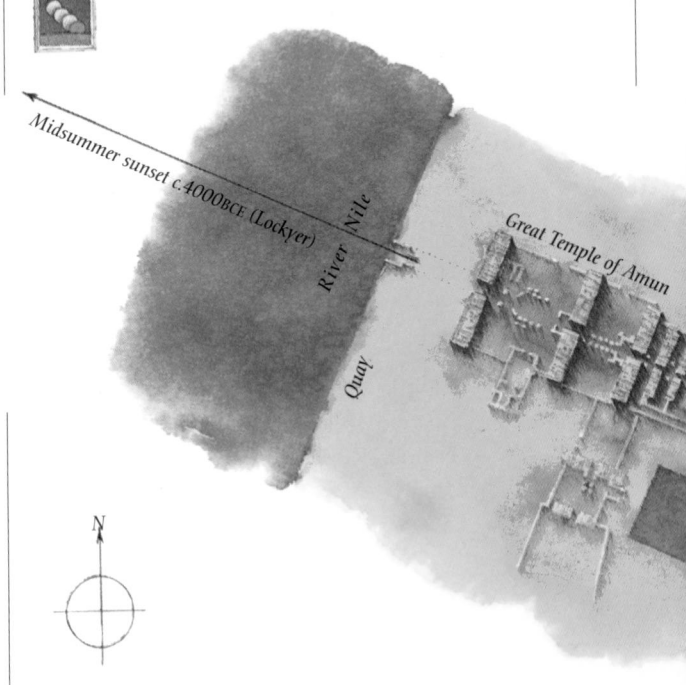

Midsummer sunset c.4000BCE (Lockyer)

River Nile

Quay

Great Temple of Amun

N

Karnak

studied this powerful axial line toward the end of the 19th century, and concluded that it was directed westward across the Nile to the midsummer sunset in *c*.4000BCE. He visualized the setting Sun shining into the the sanctuary, perhaps illuminating the sacred statue of Amun that once stood at its heart. However, this theory was disproved in 1891 by a military engineer, P. Wakefield, who saw that the view of the setting Sun would have been be blocked by the Theban Hills. Then F. S. Richards discovered in 1921 that the line was directed too far north for midsummer sunset *c*.4000BCE and could only have worked at the absurdly early date of 11,700BCE.

The indefatigable Smithsonian Institution astronomer Gerald Hawkins felt that the axis had probably been used in the other direction. He found that the

Hawkins found that Karnak was aligned to the midwinter sunrise during the Middle and New Kingdoms – not, as Lockyer had claimed, to the midsummer sunset *c*.4000BCE.

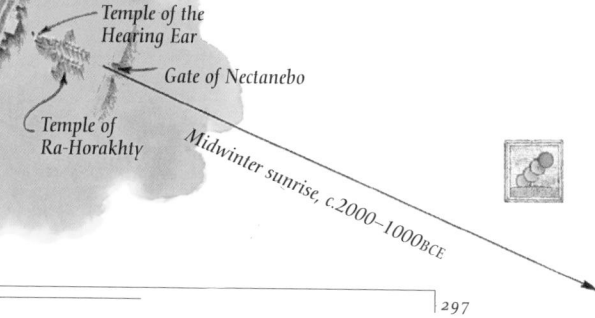

Hall of Festivals

Temple of the Hearing Ear

Gate of Nectanebo

Temple of Ra-Horakhty

Midwinter sunrise, c.2000–1000BCE

Karnak

"Hall of Festivals", built by Thutmose III (*c*.1479–1425 BCE) back-to-back with the eastern end of the sanctuary, blocked the line of sight eastward, and that this was clearly why Lockyer had looked west. But Hawkins

Amenhotep III, a king known as the "Dazzling Sun", aligned his memorial temple opposite Karnak with the midsummer sunrise. These colossi are all that remain of the temple.

found "astronomical clues" in hymns of praise "to that god that appears at dawn" in the Temple of the Hearing Ear to the east, just beyond the Hall of Festivals. Further east in the same block as the Temple of the Hearing Ear is the Temple of Ra-Horakhty, the easternmost structure on the axis before the eastern gate (the Gate of Nectanebo) in the Precinct wall. Ra-Horakhty ("Ra, Horus of the Horizon") is a form of the Sun-god, his name referring to his rising at dawn. Recalculating the axis, Hawkins found that its eastward alignment was to the midwinter sunrise between *c.*2000BCE and *c.*1000BCE. Visual obstruction was still a problem, but the answer was found in an upper chapel of the sanctuary complex, called the High Room of the Sun, where an alabaster altar, in front of a square aperture in the wall, dedicated the roof temple to Ra-Horakhty. Hawkins found that the aperture allowed a clear view of the skyline.

In 1989, the North American astronomer Ronald Lane Reese photographed the midwinter sunrise along the axis from the western entrance to the temple of Amun. He watched as the Sun "stood" over the Gate of Nectanebo for a few minutes after sunrise. He was the sole witness to the event – popular awareness has yet to catch up with this astronomical secret of Amun.

Karnak

An image of the circular zodiac found on a chapel ceiling at Dendera. The zodiacal constellations are picked out in green. They are in the order that we know them, from the first sign (Aries, on the right) moving in a clockwise direction.

DENDERA
EGYPT

The temple of Dendera, 30 miles (50km) north of Luxor, was the home of the goddess Hathor. Her well-preserved main temple is relatively late, dating from the Ptolemaic period (305–30BCE), but doubtless closely resembles earlier temples that once stood on the site.

Inscriptions at Dendera and other Egyptian temples refer to a foundation ceremony, in which the king stretched a cord between two poles to align the foundations with a particular star or constellation. At Dendera, the cord was aligned to the Bull's Foreleg (the Plough or Big Dipper).

In the 1890s, Sir Norman Lockyer became intrigued by the remains of a temple of Isis immediately to the south of the great temple of Hathor. He calculated that *c*.700BCE the axis of the main temple had been oriented to the heliacal rising of Sirius, the brightest star in the sky, and equated with the goddess Isis (see pages 160–161). The two powerful goddesses became so closely related that eventually they were effectively viewed as a single deity, "Isis-Hathor".

Dendera boasts several astrological features, including, most famously, a circular zodiac on one of the eastern roof chapels – the only one found in ancient Egypt. It is of a type developed by the Babylonians. A less familiar version in an aisle of the main temple shows the sky-goddess Nut arching over an unusual sequence of zodiacal signs – a testimony to the great magical and symbolic role that was played by the skies in the imagination of the Egyptians.

The stretching of the cord ritual at Dendera, from a mural. The goddess (left) and the pharaoh (right) wear headdresses bearing stellar and solar motifs.

HASHIHAKA

JAPAN

Astronomical orientations have been discovered at various ancient sites near the holy mountain of Miwayama, which rises above the fringe of the Yamato plain in central Japan. These sites include the

Hashihaka mounded tomb, less than a mile from Miwayama's northwest slope.

Constructed in the "keyhole" style, so called because of its groundplan, this imperial tomb probably dates from the Kofun period (300–700CE), although it has not yet been fully excavated. It has a total southwest-northeast length of 886 feet (270m). The round part of the "keyhole" rises 89 feet (27m) above the surrounding flat terrain and probably con-

An aerial photograph of the remarkable "keyhole" mound at Hashihaka – an imperial burial site embodying alignments to the midsummer and midwinter sunrises. The trees cover a structure built largely of stones and pebbles.

tains the burial chamber. Built of stones and pebbles, the monument is now overgrown with trees.

A pond or a moat once encircled most of the monument, but only two large ponds remain, the more northerly of which contains a probably artificial islet in its northwest corner. Stephan Bumbacher found that from the islet you can see a distinctive protuberance on Miwayama directly over the top of the burial mound, marking the precise point on the skyline where the midwinter sunrise occurs. He also found an alignment with the midsummer sunrise: following the axis of the monument directly over the top of the round mound from a platform on the front section, your eyes meet the summit of a large hill nearly two miles (3km) to the northeast.

Bumbacher suggests that Hashihaka's orientations could embody a perception that the spirit of the dead ruler was instrumental in maintaining the cycle of the seasons.

Hashihaka

VIJAYANAGARA
INDIA

Founded in the 14th century, the city of
Vijayanagara, about 200 miles (320km)
southwest of Hyderabad, was one of the
largest cities in the world in its day. It
was laid out as a *mandala* – a diagram
of the cosmos used as a meditation aid
in Hindu and Buddhist traditions, and
also as a groundplan for Hindu temples.
These could symbolize Mount Meru, the
mythical cosmic mountain at the centre
of the world, above which shone the
Pole Star, centre of the heavenly sphere.

John M. Fritz has suggested that the
whole royal city of Vijayanagara can be
understood as a sacred site, containing
the same mandalic spatial relationships
as the classic Hindu temple. With the
astrophysicist John McKim Malville, he
studied more than 150 temples and
shrines within the city, and found that
while many aligned north-south, some
aligned to local hills which had been

Vijayanagara

Tiruvengalanatha temple: one component in the vast cosmic diagram that is the city of Vijayanagara.

worked into the cosmological pattern. Looking north one night from a ceremonial gateway along Vijayanagara's distinctive north-south axis, Malville saw these elements powerfully synthesized. From this viewpoint, the Pole Star was seen to shine directly above Virabhadra Temple on the summit of Matanga Hill.

GAO CHENG ZHEN
CHINA

Gao Cheng Zhen

In China during the Eastern Han era (25–220CE), the town of Gao Cheng Zhen, southeast of Luoyang, represented the world axis. Ancient texts suggest that a gnomon (a shadow-throwing device) was installed there 2,000 or more years ago. Shadows vary in length at different times of the year and at different places on the Earth, and can be used to measure both time and distance.

In 725CE the Buddhist monk Yi Xing put a gnomon 8 feet (2.5m) high at Gao Cheng Zhen, one of a series on a north-south line almost 2,200 miles (3,540km) in length. Measurement of the gnomon shadows showed variations sufficient for geographical distances to be calculated. In the 13th century, the astronomer Guo Shoujing built a more sophisticated device consisting of a pyramidal tower. Projecting from its north face is a long,

The pyramidal tower at Gao Cheng Zhen is 40 feet (12m) high with a 120-foot (37m) low wall, the Sky Measuring Scale, used to make precise calendrical calculations.

Sky-Measuring Scale

low calibrated wall, the "Sky-Measuring Scale". The shadow cast by the tower was measured on the wall at noon on the summer solstice, and again at noon on the winter solstice. Measurement of the different lengths of these shadows enabled Guo to estimate the length of the year with great precision.

Gao Cheng Zhen

Sun at noon, midsummer

Sun at noon, midwinter

CAHOKIA
USA

Cahokia, a Mississippian-culture city and ritual complex in southern Illinois, was the largest prehistoric settlement in the United States. It was occupied *c*.700–1500CE and at its height covered 6 square miles (15.5km²) and had a population of 20,000. The remains cover 3.5 square miles (9km²), with 68 mounds including North America's tallest prehistoric earthwork, Monks Mound.

Originally, there were more than 120 mounds, consisting of three types: platform mounds with flat tops that supported buildings such as temples; conical mounds used mainly for burials; and ridge-topped mounds, used for burials but primarily geographic markers. Five of the eight ridge-tops fix the extreme limits of the mound area, and three align with Monks Mound to form a meridian, a north-south line: Cahokia was laid out to the cardinal points.

Cahokia

Cahokia

Monks Mound, which rises in four terraces to more than 100 feet (30.5m) was the hub of Cahokia's ceremonial landscape, the centre of the four directions. On the southwest corner of the first terrace, a temple stood precisely on Cahokia's north-south line.

Monks Mound in Cahokia is more than 100 feet (a little over 30m) high. This view is taken from the southeast and shows the two great ceremonial stairways.

Mound 72 is particularly interesting, as excavations have revealed that a huge pole some 3 feet (90cm) in diameter once stood at the exact point where the meridian passes through. A succession of mounds on the same spot contained the burials of more than 250 skeletons. The main burial, that of a 45-year-old man, was surrounded by rich grave goods: it was obviously a place of special significance, perhaps playing a key part in some kind of cosmic symbolism.

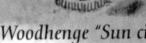

Equinox sunset

Woodhenge "Sun c[...]

On the summit of Monks Mound, the Cahokians erected a huge timber building, presumably a ruler's dwelling or a temple. A massive pole outside this building may have been a Native American version of the World Tree.

In 1961, the archaeologist Warren Wittry and his team found evidence to the east of Monks Mound, just south of the east-west line, of a succession of five timber circles ("woodhenges"), which he carbon-dated to a period *c.*1000CE. The third circle in the sequence, 410 feet (125m) across, and now partially reconstructed, had poles marking three key sunrise positions – midsummer, equinox and midwinter. This Sun circle also had symbolic value: not only did a pole mark the equinoctial sunrise as viewed from the Sun-watcher's pole, but so did Monks Mound on the skyline. With potent symbolism, perhaps during some form of ritual linking the Sun with the ruler, the Sun would have appeared to rise out of the building atop the mound.

The pattern on a beaker found at the midwinter sunrise marker post in Cahokia's "Sun circle" is thought to depict the circle's solar alignments.

Cahokia

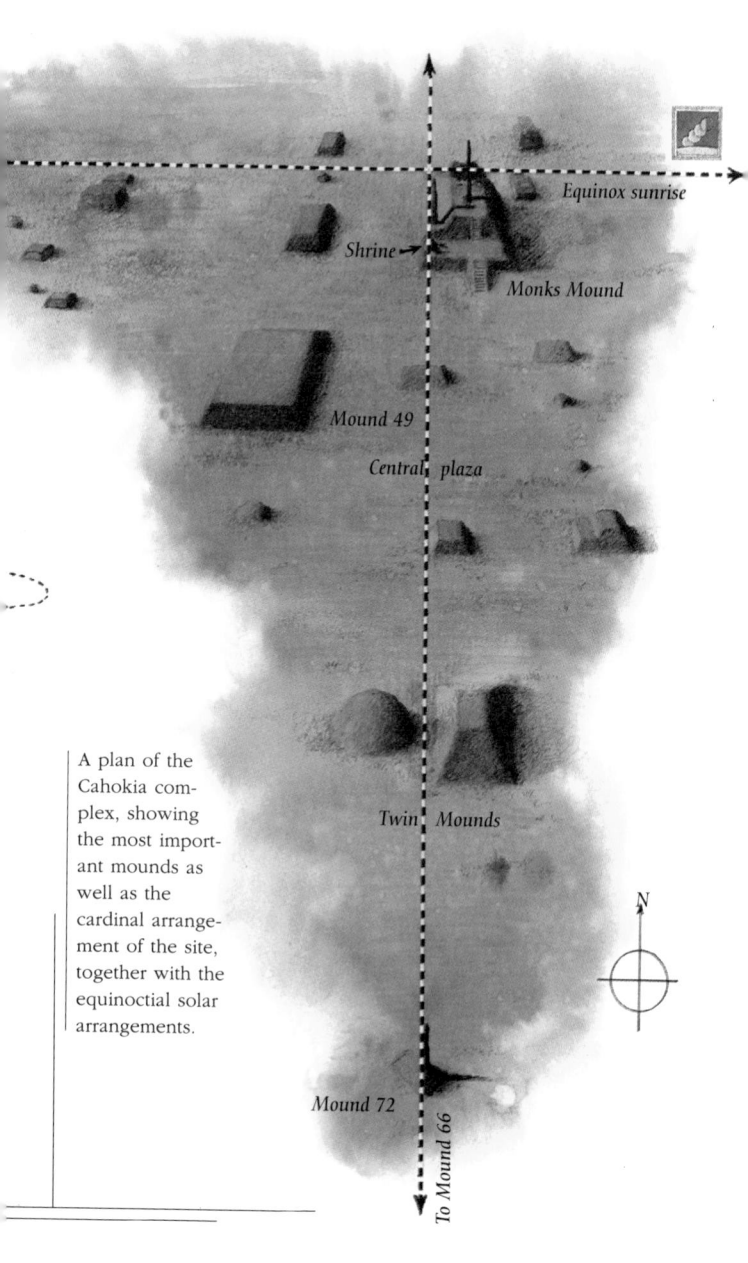

Equinox sunrise

Shrine →

Monks Mound

Mound 49

Central plaza

A plan of the
Cahokia com-
plex, showing
the most import-
ant mounds as
well as the
cardinal arrange-
ment of the site,
together with the
equinoctial solar
arrangements.

Twin Mounds

N

Mound 72

To Mound 66

Casa Grande

CASA GRANDE
USA

Casa Grande ("Big House") is a mysterious four-floor structure standing in an otherwise almost vanished Hohokam-culture village in southern Arizona. The Hohokam ("Those Who Have Gone") appeared in the area *c*.300BCE, probably from Mexico, and were active up until the 14th century CE. Little is known about them, but they were accomplished canal engineers, using irrigation to farm in the desert. They were also fine jewelers, and seem to have lived peacefully. Their culture reached its height between 1000CE and 1400CE.

Although Casa Grande was abandoned by 1450, enough survives today for us to appreciate its structure. Some 35 feet (10.7m) high on a base of 60 feet by 40 feet (18m by 12m), its walls of sunbaked mud are nearly 6 feet (2m) thick at their bases, tapering to 2 feet (70cm) at full height. The building had over 600 roof beams of pine, juniper and fir, imported from more than 50 miles (80km) away. The whole structure contained 11 rooms, with a top floor consisting of a single room.

Casa Grande has two intriguing openings in the upper corners of its west wall, one circular and one square. The round aperture gives a sightline to the midsummer sunset, while the square one opens to an extreme setting point of the Moon during its 18.6-year cycle. The top floor also has openings aligned to significant sun- and moonrises.

Most authorities believe that Casa Grande was some kind of observatory,

Casa Grande's square aperture (upper right) aligns to the Moon's most southerly setting point. Its round counterpart (upper left) is aligned to sunset at midsummer.

Casa Grande

Casa Grande

but a clue to its deeper significance may lie in an observation of 1887 by the anthropologist Frank H. Cushing, who had lived among nearby Pueblo Indians. Cushing noted that the building's floor-plan resembled the pattern laid out in the ceremonies of neighbouring Hopi peoples for consecrating the cornfields.

This insight was developed by another anthropologist, David Wilcox, who proposed that Casa Grande's astronomical function expressed concepts enshrined in cornfield consecration ceremonies. According to these, the central area of a cornfield represents the "Hill of the Middle", with the surrounding areas of the field representing the "Hills" of the four cardinal directions, plus the zenith and the nadir. According to this theory, Casa Grande is a three-dimensional realization of the ceremonial pattern: the central tier of rooms represent the nadir, zenith and "Middle" or World Centre; the rooms around represent the four directions.

CHACO CANYON

USA

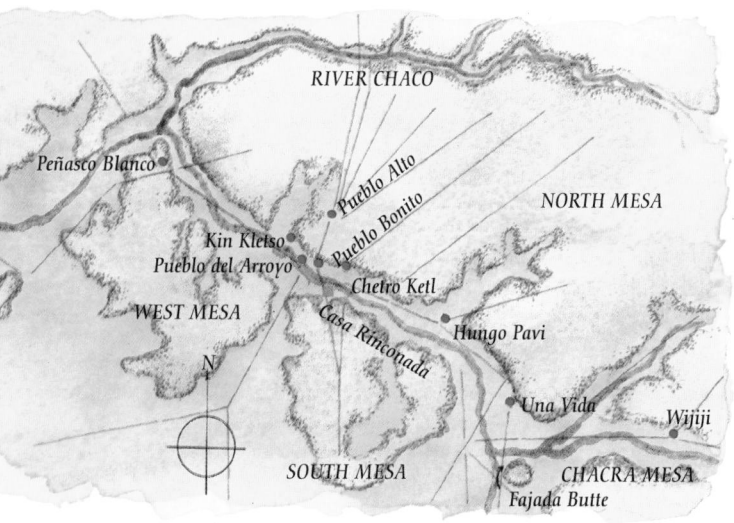

A map of Chaco Canyon showing the nine Great Houses. The brown lines are the mysterious "roads" that radiate for many miles from the canyon.

Chaco Canyon was inhabited by the Anasazi people (Navajo for "Ancient Ones") – a thousand or more years ago. The Anasazi first appeared in the later centuries BCE, probably from Mexico, and reached their cultural height between 900CE and 1300CE. They were not only great builders, engineers and traders, but also seemed to have been adept at ceremonial astronomy. The culture collapsed in the 14th century; the Pueblo peoples of the southwestern US are their successors.

The canyon is a broad, shallow sandstone gorge cutting east-west across the arid Chaco Plateau in New Mexico. Archaeologists now think that Chaco, though possessing a permanent population, was primarily a ritual centre that attracted large numbers of people from the region at certain times of the year.

Scattered along the canyon floor are the ruins of nine "Great Houses", multi-storied "villages" with walls, courtyards,

Chaco Canyon

A view of Pueblo Bonito, the largest of the Great Houses in Chaco Canyon. In the foreground are two great *kivas* – circular chambers built for ceremonial purposes. The complex is laid out on a strictly cardinal plan, with a low meridional (north-south) wall dividing it into eastern and western sectors.

storage pits and *kivas* – large ritual chambers. The Great Houses seem to have been essentially ceremonial buildings. One of them, Pueblo Bonito, said to be the largest prehistoric ruin in the USA, covered three acres (1.2ha) and contained 800 rooms, two great *kivas* and 37 smaller ones. All the Chaco Great Houses were built between 900CE and 1115CE. A further 150 have been identified all over the canyon.

At various locations along the canyon walls are rock paintings and carvings, some Navajo and others Anasazi, and these are thought to have marked positions for Sun-watcher priests. In relatively recent times, and still today to an extent, Pueblo peoples such as the Hopi and Zuni had a tradition of Sun priests who watched the rising Sun against the local skyline throughout the year. When

Chaco Canyon

The leaning rock slabs at Fajada Butte, which channel shafts of sunlight onto spiral designs carved on the outcrop behind, in a deliberate contrivance of solar symbolism.

sunrise appeared at certain significant points on the horizon, the Sun priest would announce that specific ceremonies or the planting of particular crops were impending. A white-painted symbol with rays streaming out from four sides may be seen near Wijiji Pueblo, the easternmost ruin in Chaco Canyon. We may presume that this is a Sun symbol, since a person standing near it is able to witness the midwinter Sun rising from behind a natural sandstone pillar on the canyon rim opposite.

On a high ledge near the summit of Fajada Butte, a 430-foot (130m) sandstone outcrop at the southeastern entrance to the canyon, three fallen slabs lean against the rock wall. One late morning

Chaco Canyon

shortly after the summer solstice in 1977, the artist Anna Sofaer observed a sliver of sunlight suddenly project through a gap between the slabs onto the shaded area of the rock wall behind. On that wall she found two spiral rock carvings. The sliver of sunlight was cutting almost through the centre of the spiral design. Subsequent study of this site showed that at noon on the winter solstice, when the Sun shines at a lower angle in the sky, two slivers of light project between the slabs, perfectly framing the larger spiral carving. At the equinoxes, one long sliver of sunlight cuts the large spiral just to one side of its centre, while a shorter sliver bisects the smaller carving. There could hardly be stronger evidence of an Anasazi solar calendar.

At Pueblo Bonito, as midwinter approaches, a slice of sunlight flashes

Two details showing how the spiral designs carved on an outcrop of Fajada Butte (see illustration on opposite page) are traversed by slivers of sunlight at the solstices. Similar spiral motifs represent the Sun in numerous cultures across the world.

Midsummer, noon

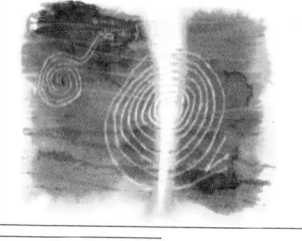

Midwinter, noon

Chaco Canyon

through an unusual corner aperture in an upper room within the complex, and is projected onto a wall. As the days progress, this sliver of sunlight broadens until by the morning of the solstice it casts a rectangle of light.

The very structures of Casa Rinconada and Pueblo Bonito, directly across the canyon floor from one another, speak of astronomical awareness. Casa Rinconada is aligned to the four cardinal directions, its two large, T-shaped entrances precisely aligned north-south. This alignment also gives a sightline to Pueblo Alto, a ruined Great House high on the north rim of the canyon. Pueblo Bonito was similarly laid out on a cardinal plan, divided into western and eastern halves by a low wall that lies on a virtually exact north-south meridian.

Vestiges of ceremonial astronomy among today's Pueblo peoples lend valuable clues to the solar architecture and symbolism of a mysterious, but undeniably great, lost civilization.

HOVENWEEP
USA

The eastern room of Hovenweep Castle. Light from the midsummer and midwinter sunsets enters through apertures in the west and south walls, and the setting equinoctial Sun enters through the south doorway.

Consisting of six groups of Anasazi (see page 315) buildings approximately 800 years old, Hovenweep National Monument straddles the border of southeast Utah and southwest Colorado. The sites, in and around the edges of small box canyons, were not documented until 1874, and the buildings are today much as the Anasazi people left them, except for the depredations of time. There are towers on the canyon floors, next to precious water sources, and clusters of other buildings perched on the edges of canyons or even on top of rock outcrops. Some doorways open onto empty space, and were presumably reached by rope ladders.

At a group of D-plan towers and

Midsummer sunset

Equinox sunset

Midwinter sunset

Plan

Hovenweep

walls called Hovenweep Castle, some of which rise to 20 feet (6.1m) in height, the archeoastronomer Ray Williamson suggested one possible set of astronomical events. The tower on the western side contains a room with two small apertures in the walls aligning respectively to the summer and winter solstice sunsets, while the outer and inner doors line up to the equinox sunsets. The chances of all three alignments occurring by chance are one in 216,000, so we may assume that they were deliberately created.

Further along the canyon are the ruins of what archaeologists call Unit-Type House. Its east room has four slits, two aligned to the midsummer and midwinter sunrises respectively, and another oriented to the equinoctial sunrises. A wall inside this room is so angled that it presents a perfect surface to receive the sunbeams through each of these apertures on the relevant days. Williamson has also noted possible astronomical events at other Hovenweep sites.

TEOTIHUACÁN
MEXICO

A great urban and ceremonial centre about 30 miles (50km) northeast of Mexico City, Teotihuacán was laid out in the 1st century CE, reaching its cultural peak between 350CE and 650CE. At that time it had a population of some 200,000 and extended to around 10 square miles (26km²). Consisting of temples, shrines, plazas, dwellings and workshops, it was larger than any city in Europe. By the 8th century, however, it had been destroyed by fire and abandoned.

No one knows who the builders of Teotihuacán were, but their great religious and economic centre dominated the Valley of Mexico and beyond for more than 1,000 years prior to the Aztecs, who encountered its awesome ruins and considered it the source of all civilization. It was they who named it Teotihuacán, "Birthplace of the Gods". In Aztec mythology, it was where

Teotihuacán

Nanahuatzin, a dying god, jumped into a ceremonial fire which the four creator gods (representing the cardinal directions) were too fearful to enter. Turned to flame, Nanahuatzin became the "Fifth Sun", the Sun of the present cosmic age. His companion Tecciztecatl joined him in the fire and became the Moon. The Fifth Sun agreed to orient the world by his rising and to organize the passage of time. The Aztecs decided that the larger of Teotihuacán's two great pyramids was dedicated to the Sun, the smaller to the Moon.

Teotihuacán was laid out to a four-part plan, but, interestingly, it was not primarily oriented to the four cardinal directions. Instead, it was organized to an angle of 15.5° east of north. This originally puzzled archaeologists and astronomers, who observed that this orientation had been strictly applied throughout by the city's builders – even

A view of the Pyramid of the Moon at Teotihuacán, showing the courtyard in front of it and the end of the Street of the Dead. The original surrounding buildings are now in ruins.

the San Juan River had been canalized to conform. This skewed meridional axis is marked by the great ceremonial way which the Aztecs called the Street of the Dead. The street extends for 1.5 miles (2.4km), aligning with the Pyramid of the Moon at its northern end (and to a notch in Cerro Gordo, the city's sacred mountain and water source). The Pyramid of the Sun is just to the east, its sides oriented in parallel with the smaller companion structure.

The tiered bulk of the Pyramid of the Sun rises to well over 200 feet (60m). Its western side faces toward Cerro Colorado and Cerro Maravillas, both of which may have been regarded as sacred to the Teotihuacános. However, researchers have sought a primarily astronomical answer to the mystery of the city's odd orientation.

It is now realized that the site of Teotihuacán originated as a result of the existence and ritual use of a cave lying beneath the Pyramid of the Sun. The

Teotihuacán

Teotihuacán

cave consisted of four lobes with an extension originally created by a flow of volcanic lava. In plan it resembled a four-leafed clover with a long stem. Archaeologists have found clear evidence of ritual and ceremony inside the cave. The cosmological significance of the four lobes would have been apparent to the Mesoamerican builders of Teotihuacán, because they, like all indigenous Mesoamericans, had a conception of the universe based on the four cardinal directions. They also granted great significance to the intercardinal directions – northeast, southeast, southwest, northwest. The cave had even more to offer: by a remarkable coincidence, the lava tube passage aligns to the setting point of the Pleiades, which had great symbolic import for ancient Mesoamericans. The first pre-sunrise annual appearance of the Pleiades heralded the first of the two occasions

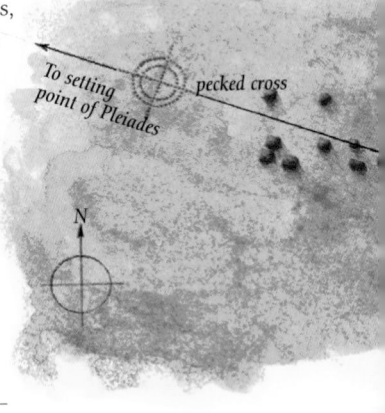

To setting point of Pleiades

pecked cross

N

each year when the Sun passes directly overhead, through the zenith, at the latitude of Teotihuacán. At noon on these days, shadows dwindle to nothing, and it was believed that the Sun-god visited the Earth briefly at these moments. Furthermore, it so happens that at the latitude of Teotihuacán the Pleiades themselves pass close to the zenith.

The Pyramid of the Sun was built directly over the cave and was oriented along the line of the lava tube passage, thus also aligning to the point where the Pleiades set in 150CE. Also in the alignment were the mountain peaks toward the Pleiades' setting point. The American archaeoastronomer Anthony F. Aveni realized that the skewed north-south axis of Teotihuacán is precisely perpendicular to the east-west Pleiades axis.

A plan of Teotihuacán showing the city's alignment to the setting point of the Pleiades *c.*150CE, and two of the "pecked cross" patterns that reflect the overall orientation of the site.

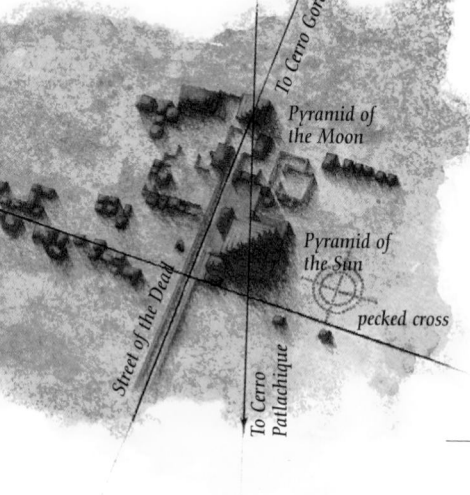

To Cerro Gordo

Pyramid of the Moon

Pyramid of the Sun

pecked cross

Street of the Dead

To Cerro Patlachique

Teotihuacán

Teotihuacán

The east-west axis crosses the Street
of the Dead just south of the Pyramid of
the Sun. Nearby, the first of more than
30 pecked crosses at the site was found.
About a metre across, these consist of
two concentric rings of holes crossed by
two lines, pecked into the floors of
buildings or rocks. The crosses' axis
aligns with the city's grid. "Teotihuacán
north" clearly had precedence over
astronomical north, but the pyramids of
the Sun and Moon form a regular north-
south alignment and some other buildings acknowledge this orientation.

A cross-section
of the Pyramid
of the Sun. The
plan shows the
lava-tube cave
and "clover-leaf"
chamber. The
cave was partly
straightened to
align with the
setting point
of the Pleiades.

Cave   E L E V A T I O N

*Straightened
Passage*   ✴ PLAN

Uxmal
MEXICO

The ruined city of Uxmal, in western Yucatán, about 50 miles (80km) south of Mérida, is a major architectural site of the Maya culture. The complex flourished *c.*600CE–900CE, but it had been occupied at least as far back as the early centuries CE.

A particularly intriguing structure at Uxmal is the "Dwarfs' House". Built on the summit of a pyramid, it is far too small for human occupants, and many observers believe that this and similar structures elsewhere were intended for the use of spirits or supernatural beings.

Most of the buildings in the city share a roughly cardinal orientation (actually 9° east of north). However, the rectangular, 322-feet (98.2m) long Palace of the Governor, an impressive monument clad in a mosaic of some 20,000 sculptured stones, is noticeably skewed from this: its long axis is turned 19°

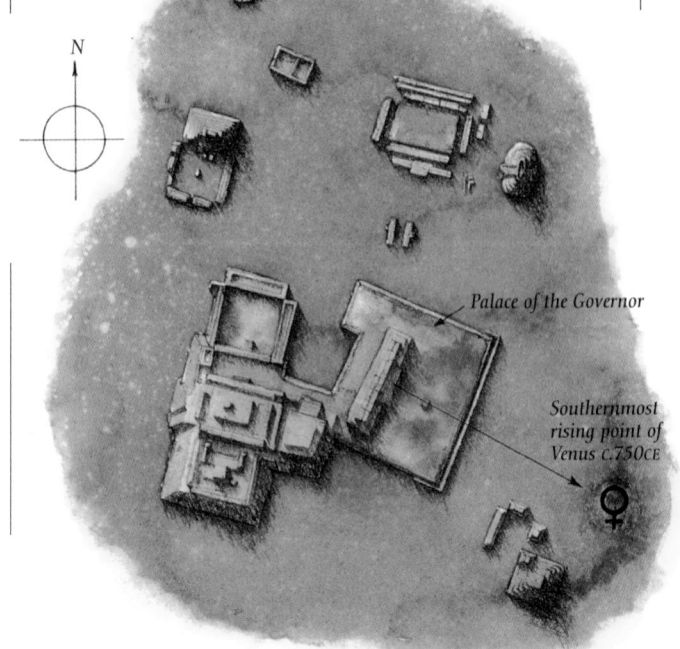

Uxmal

Palace of the Governor

Southernmost rising point of Venus c.750CE

♀

A plan of Uxmal shows how the Palace of the Governor is on a different orientation from the rest of the site: it points to the southernmost rising point of Venus.

clockwise from the usual axis, so that its frontage faces southeast. It sits on a stepped platform on a natural elevation. A sightline directly out of the central doorway in the building's façade passes over a cylindrical carved stone block and a platform supporting a double-headed jaguar sculpture to a pyramid on the skyline a few miles away. This was identified by Anthony F. Aveni as part of Nohpat, another ruined Maya centre. Viewed from the central entrance of the Palace, this marks the southernmost rising point of Venus.

Venus was a major player in Mayan cosmology, one of the Twins in the epic Mayan poem, *Popol Vuh* (the other Twin being the Sun), associated with warfare and death. Sacrificial victims were sometimes painted blue, the symbolic colour for Venus, prior to death. The association between Venus and Uxmal's Palace of the Governor was further emphasized by hundreds of stone masks of the god Chac along the upper façade: the lower eyelids of each mask are carved with a glyph representing Venus.

Maya priests had almanacs recording the cycles of Venus, which were keyed into their overall calendar, so that they could identify propitious times for ritual combat and sacrifice. This was a complicated business. Five 584-day Venus

Uxmal

Uxmal

cycles fortuitously equal eight 365-day solar cycles, so an eight-year Venus almanac was devised. But that, in turn, had to mesh with the Maya's 260-day sacred almanac, which involved a cycle combining a sequence of 13 sacred numbers with another of 20 named days. The result of this computation was a 104-year Great Venus Almanac, which incorporated 65 Venus cycles and 146 sacred almanacs!

Interestingly, the Uxmal-Nohpat sightline is marked on the ground by a straight sacred way, a *sacbe* ("white road"). These causeways – most are now very overgrown – link numerous ancient Maya cities. Some researchers believe that the significant astronomical alignments at Mayan centres were in effect transcribed onto the ground on an enormous, territorial scale.

The double-headed jaguar sculpture in the courtyard of the Palace of the Governor. It falls on a sightline from the centre of the Palace doorway to the southern rising point of Venus.

CHICHÉN ITZÁ
MEXICO

Chichén Itzá, in the Yucatán Peninsula, was an important ceremonial centre during both the Maya and Toltec periods. "Old Chichén" was built by the Maya *c.*600–830CE. According to Anthony F. Aveni, the Caracol (Spanish for "Snail") at Chichén Itzá – so called because of its internal winding staircase – is one of "the most secure examples of the incorporation of a horizon-based astronomy in architecture". It consists of a two-tiered rectangular platform supporting a cylindrical tower, at the top of which are three surviving "windows", really horizontal shafts, that open westward.

Chichén Itzá

The pyramid El Castillo at Chichén Itzá (see also page 336).

As long ago as the 1920s, the North American archaeologist Oliver Ricketson suggested that the Caracol's openings could have been used for astronomical sightings. The

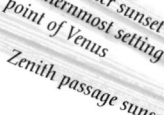

surrounding terrain is flat, with no distant hills or peaks to act as foresights for sightlines, but Ricketson proposed that accurate sightings might have been achieved by looking out of these shafts diagonally – for example, from the inside right to outside left jamb of the opening. This gives a "gun-sight" method of lining up two points on a specific place on the distant horizon.

Used in this way, the shafts do yield significant astronomical events. The

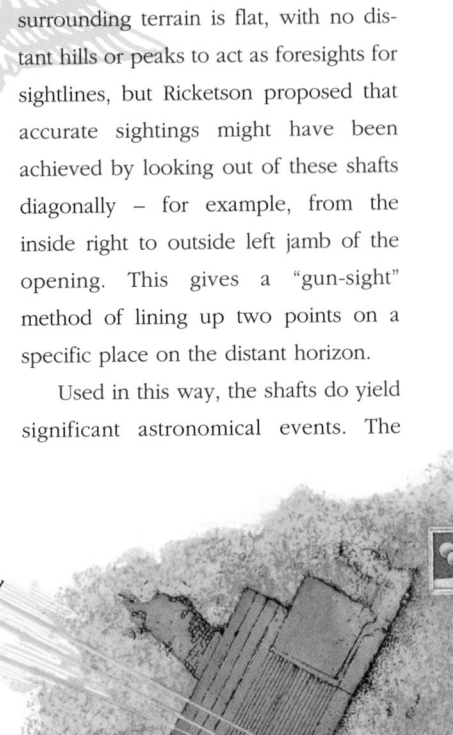

Midsummer sunset and northernmost setting- point of Venus

Zenith passage sunset

Midwinter sunset

The Caracol, showing its astronomical orientations. The main plan shows the square platform on which the round tower stands, with descending stairways to the west. The inset shows the room in the upper part of the tower.

sighting lines in Window 1 and the narrower Window 2, have been found to frame perfectly the most northerly and southerly setting points of Venus, which in Maya cosmology represented the plumed serpent god Kukulcan, the equivalent of the Toltec and Aztec Quetzalcoatl. Moreover, the inside-right-to-outside-left diagonal of Window 1 aligns to the equinox sunset, framing a tiny wedge of sky into which the setting equinox Sun fits perfectly.

The tower stands on a platform, one diagonal of which aligns to the midwinter sunset in one direction and the midsummer sunrise in the other. A niche containing a pair of columns in the platform's main stairway is skewed at an angle to the upper plat-

Chichén Itzá

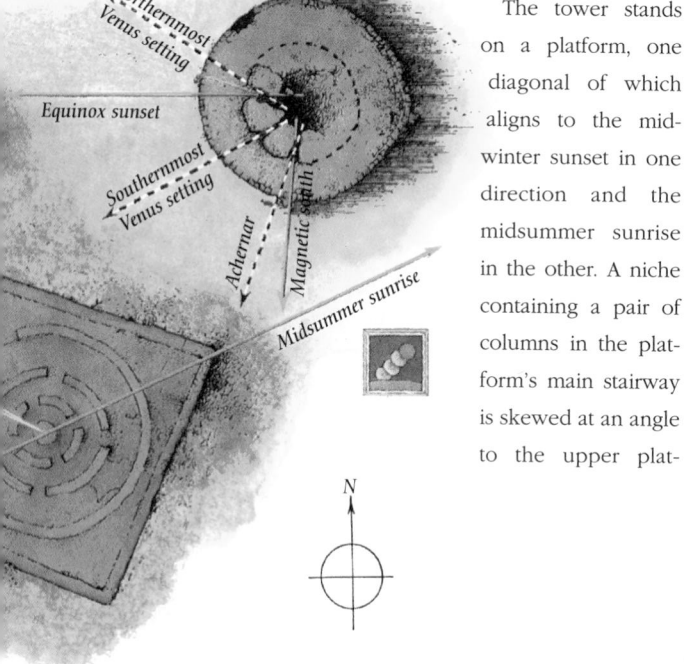

Northernmost Venus setting

Equinox sunset

Southernmost Venus setting

Achernar

Magnetic south

Midsummer sunrise

N

Chichén Itzá

A plan of El Castillo (see also page 333) showing its alignments to the solstices. In addition, the equinox sunsets create a spectacular "light show", as patterns of light and shade move down the west-facing balustrade of the north stairway.

form, and Aveni discovered that this was oriented on the northern Venus extreme.

Astronomical alignments are also found at the stepped pyramid of El Castillo ("The Castle"), known also as the Temple of Kukulcan, about half a mile (800m) northeast of the Caracol. Some 75 feet (23m) in height, it has a 91-step stairway up each of its four faces. The four lots of 91 steps plus the top platform, effectively a top step shared by all the stairways, gives 365 steps. Other Mayan structures also contain 365 features, and archaeologists are confident that this is a deliberate reference to the solar year.

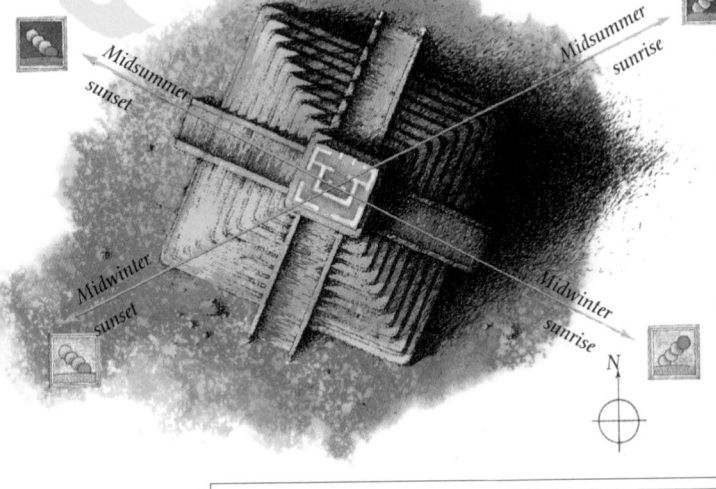

Midsummer sunset

Midsummer sunrise

Midwinter sunset

Midwinter sunrise

N

MACHU PICCHU
PERU

The spectacular remains of Machu Picchu are some 60 miles (100km) north of Cuzco. Because of its position on a mountain saddle 8,000 feet (2,440m) aloft in the Andes, this Inca citadel was never discovered by the Spanish. The site consists of cultivation terraces, stone houses, temples, plazas and residential compounds clinging to a ridge between two peaks, Machu (old) Picchu and

A general view of Machu Picchu, an Inca citadel precariously balanced between the jungle and the clouds. The Inca gods included the Sun-god Inti, the supreme creator Viracocha and the Moon-goddess Mama Kilya.

Huayna (new) Picchu. The citadel, which is reached by a stairway of more than 3,000 steps, was seen for the first time since its abandonment by Hiram Bingham of Yale University in 1911.

Archaeoastronomers have found two interesting features at Machu Picchu. The most important of these is the Intihuana, the "Hitching-Post of the Sun". The major Inca festival of Inti Raymi (Inti was the Sun-god) took place at the winter solstice (June 21 in the Southern Hemisphere). Each year at this time, the Inca priests performed a ceremony to "tie the Sun" to prevent it from swinging ever farther north in its daily arc and be lost for ever. The Inca are known to have had other *intihuanas* in their empire, but the Spanish conquerors destroyed them as pagan symbols: the Intihuana at Machu Picchu is the only one known to have survived.

Situated on top of a natural spur, this curious object is carved out of a single piece of granite. It consists of an upright

about a foot (30cm) high emerging from an asymmetrical platform. The upright of the Intihuana is assumed to be a shadow-throwing device, a gnomon, like that on a sundial. Gerald Hawkins noted that its shadow could be read to within half an inch (1.25cm), the equivalent of a quarter of a degree. The solstices and equinoxes and even variations in the lunar cycle could all be "read" by the object.

The other point of astronomical interest at Machu Picchu is the Torreon. This now roofless rectangular temple structure has a northeast-facing window centring on the midwinter sunrise. It also opens toward the rising place of the Pleiades, while the temple's southeast window aligns to the rising stars known as Collca (the Storehouse) to the native people, in the tail of our Scorpio.

Machu Picchu

UAXACTÚN
GUATEMALA

Midsummer sunrise

Uaxactún, in Guatemala's Petén rainforest, is a Mayan ceremonial centre of the period 250–450CE. Three cross-shaped petroglyphs at the site, almost identical to those at Teotihuacán and similarly oriented (see page 328), indicate a very early foundation for the city.

In the 1920s, archaeologists began to suspect that a group of buildings consisting of a pyramid and three temples to the east of it (Group E) was a solar observatory. However, it was only in 1978 that Anthony F. Aveni showed that, if viewed from the pyramid, the equinox Sun rises over the central temple. From the same vantage point, the midsummer Sun rises along an edge of the northern temple and the midwinter Sun slides up along the edge of the southern temple. Aveni suspects that the sides of the central temple may have marked other key sunrises in the complex Mayan calendar.

A diagram of the solstice and equinox sunrise effects at Uaxactún's Group E temples. The viewing positions, from a stairway, are shown in the pyramid plan.

Uaxactún

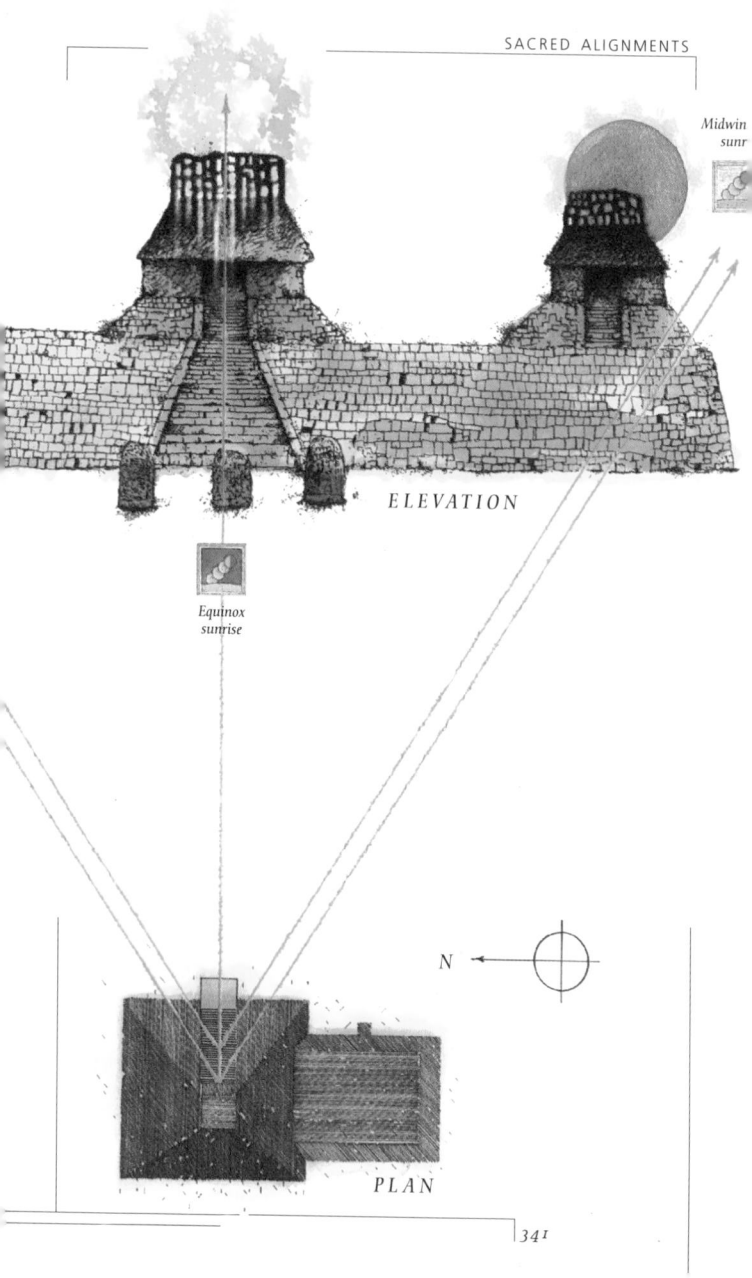

Midwin
sunr

ELEVATION

Equinox
sunrise

N ←

PLAN

CUZCO
PERU

The Inca people, who rose to supremacy in the Andes in the 14th century CE, called their capital Cuzco – the "Navel" (of the Earth). Its symbolic planning was begun in 1440, when the main buildings were laid out on a grid oriented intercardinally (southwest-northeast, northwest-southeast). This orientation derives from the observation of the Milky Way over a period of about 12 hours in which it appears to divide the sky and the horizon intercardinally.

If Cuzco was the navel of the Inca empire, then the Corincancha, the temple of the Sun, was the navel of Cuzco. It was aligned to the June (winter) solstice sunrise, when the emperor would sit in a recess decked in gold plates and precious stones. The solstitial sunbeams striking the recess made the Inca resplendent as the "Son of the Sun".

Midsummer sunset

Tower

Mid-August sunset

Tower

Equinox sunset

ABOVE A plan of Cuzco showing its division into four quarters, as well as the solar sightlines from the Ushnu – a stone pillar acting as an observatory in the main plaza.

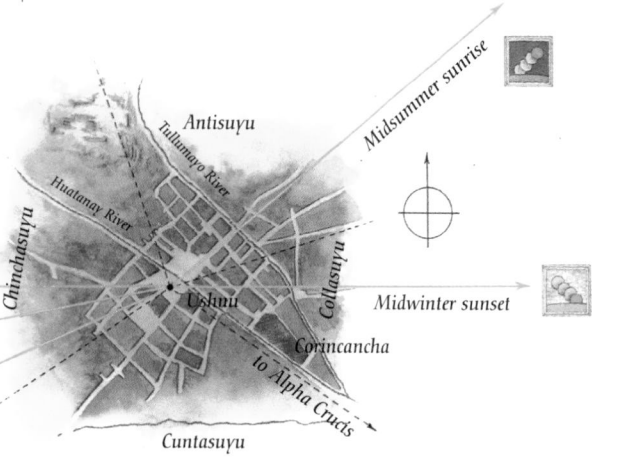

The Coricancha was the centre of 41 radiating *ceques,* alignments linking *huaca*s (sacred places and peaks). The number of *huaca*s totalled 328, the number of days in the Inca year. Those living on each *ceque* would organize the festivals at the time of year relevant to their *ceque*.

BELOW A plan showing the *ceques* (sacred alignments) radiating from Cuzco's centre.

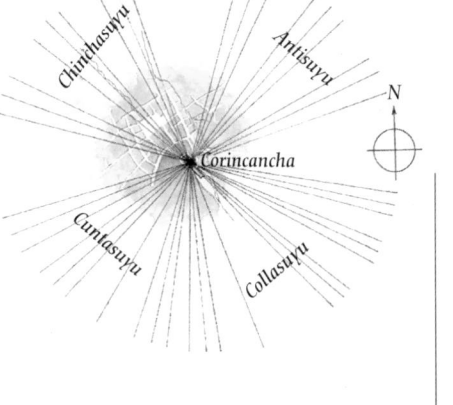

MISMINAY
PERU

The intercardinal direction system of Andean peoples has been thoroughly examined by the anthropologist Gary Urton. Concentrating his research at Misminay, a small community of Quechua descendants of the Inca, Urton found that the ancient cosmology still lives.

In the Andes the Milky Way is a very striking feature of the night sky. It girdles the Celestial Sphere, so that when half is visible, the other half is not. The plane of this ring of starlight, however, is offset by that of the Earth's rotation, so that its rising and setting patterns seem to "tumble". As it rises, it rolls up from the east, spanning the sky. When its course crosses the zenith, it stretches out in a great diagonal – say, northwest-southeast. Twelve hours later, after the Earth

BELOW The snow-clad Andes in the region of Misminay, a village 16 miles (25km) from Cuzco. Gary Urton has made a detailed study of cosmological beliefs in the Misminay area.

has turned, the band of the other hemisphere of the Milky Way passes through the zenith, its axis running northeast-southwest. Thus, the motions of the Milky Way have the effect of cutting two lines in the sky, crossing at the zenith and dividing the sky into four quarters.

The zenith point where this notional "crossing" of the two halves of the Milky Way occurs is called Cruz Calvario (the Cross of Calvary) by the people of Misminay, who mirror this cosmic scheme in the layout of their community. Two intercardinal paths cross at a point marked by a small chapel in the centre of the village. This is the Crucero (Cross). The irrigation canals in Misminay also follow this X-shaped pattern, running alongside the courses of the paths – a water pattern that has much significance for the people, to whom the Milky Way is Mayu, "River", thought to carry water from the cosmic ocean in which the world floats and to redistribute it as rain. To the inhabitants of Mis-

Misminay

NORTH

Misminay

minay, the nearby Vilcanota River, which flows from southeast to northwest, is a reflection of the Milky Way. The Cruz Calvario likewise mirrors the Crucero beneath – an invisible axis can be imagined connecting the two.

This plan, clearly a vestige of that employed by the Inca, but now using Christianized terms, underpins the whole worldview of the local people. When standing at the Crucero, for example, one can see specific features on the skyline that have mythological import for them. In particular, there are the sacred mountains or *apus*, their peaks representing the points where earth and sky meet. The northerly sector – the quadrant between the northwest and northeast arms of the intercardinal plan – is the realm of the ancestors, of the dead. The name of the main *apu* there, Wañumarka, means "Storehouse of the Dead". This sector is where the ancestors of the present-day inhabitants of Misminay are said to have first settled.

NE

Wañumarka

NW

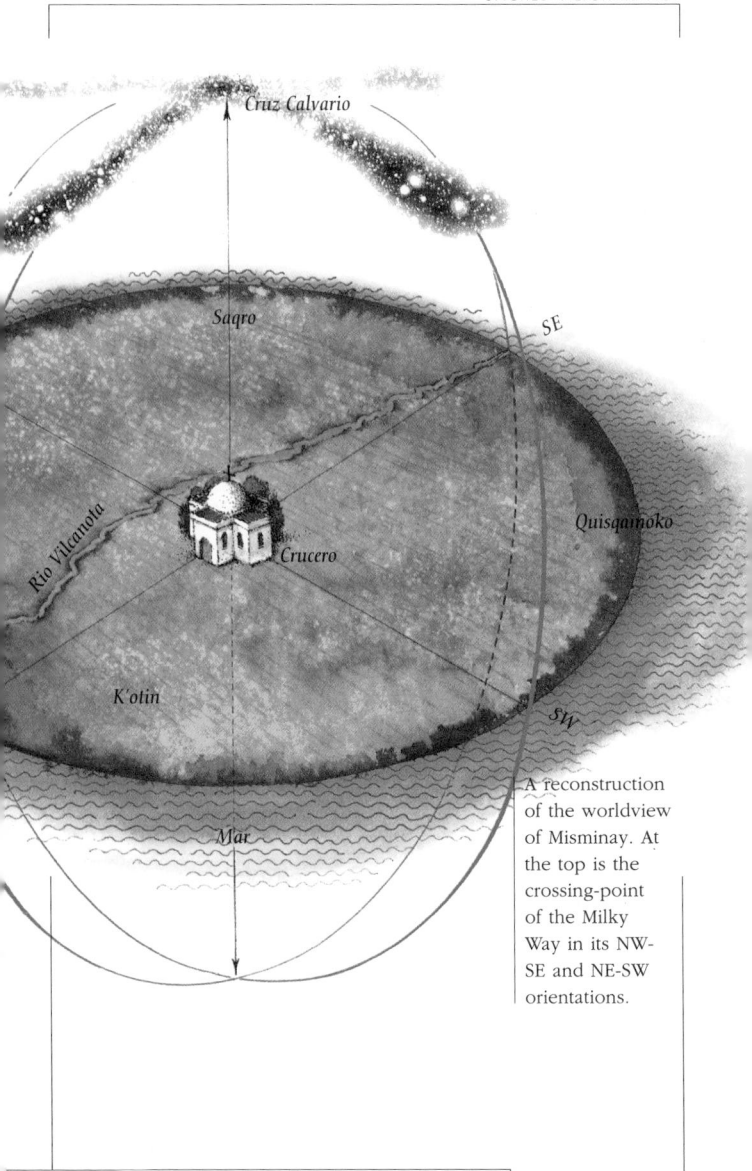

A reconstruction of the worldview of Misminay. At the top is the crossing-point of the Milky Way in its NW-SE and NE-SW orientations.

MEDIEVAL EUROPEAN SITES

The practice of aligning temples to heavenly bodies survived in Europe beyond prehistoric and pre-Christian times. This is partly an effect of the early Church's policy of Christianizing

Medieval European Sites

The 8th-century CE marble throne of Emperor Charlemagne in the octagonal chapel of Aachen Cathedral in western Germany. On midsummer's day, a sunbeam would have penetrated the chapel to spotlight the emperor's face or crown.

pagan traditions, but even today churches and graves tend to be aligned east-west, with church altars at the east, toward the rising Sun (the direction from which, according to Christian cosmology, the Second Coming will occur).

Medieval churches were often supposedly aligned to the point of sunrise on the day of the saint to whom they were dedicated, rather than to the east. A survey of almost 300 English churches by the Reverend Hugh Benson in the 1950s revealed that a significant proportion were indeed aligned to sunrise on their patron saint's day. He also found that the very ancient St Piran's church in Cornwall was aligned to a prehistoric earthwork over two miles away – the point where the Sun would have risen on August 15th in the 7th century CE.

It also seems that in early Europe astronomy was used in a more complex way, although very little documentary evidence has survived. Chartres in France is one place that contains clear

Medieval European Sites

indications of medieval sacred astronomy. The epitome of Gothic cathedrals, this magnificent building was constructed on the site of the chief Druidic centre of the Gaulish Carnutes tribe, where the Romans subsequently erected a shrine. Some scholars argue that the sanctity of the site even predated the Iron Age Celts, being marked by a Neolithic mound covering a stone chamber or dolmen.

A mysterious feature of Chartres is the fact that at noon on the summer solstice, when the Sun is at its highest, a ray of sunlight shafts through a small area of clear glass in the window dedicated to St. Apollinaire on the western side of the south transept. This shaft falls on a flagstone which differs from those around it, being larger, differently coloured, set at an angle, and, most significantly, incorporating a small, round metal disk. The solstitial sunbeam falls on this disk. Some scholars have dismissed this as coincidence, but other researchers believe that pre-Christian elements were built into the cathedral's remarkable fabric by its master masons.

Another great cathedral that may have special astronomical interest is that of Aachen, in Germany. The presence of hot springs on the site, believed to have healing properties, gave the place an importance that goes back at least to the Roman and Celtic periods. In the 8th century CE, the emperor Charlemagne built his imperial palace here. His octagonal palace chapel, sited over the old

A print showing a sunbeam piercing St Martin's Hole in the Tschingelhorner peak to fall on Elm church in the Swiss Alps.

Medieval European Sites

Roman baths, lies at the core of the present cathedral. In the late 1970s, the German photographer Hermann Weisweiler was waiting for the best light for a photograph when he witnessed a ray of sunlight suddenly flashing through an upper window of the octagonal chapel at exactly 90°. Impressed by this event, Weisweiler carried out further research which revealed the chapel to be a veritable sundial.

At noon on the summer solstice, a ray of light fell directly onto the golden ball hanging from the chapel's domed ceiling, where the "Barbarossa chandelier" depicts the heavenly Jerusalem. On the same day, a ray would also have illumined the face (or crown) of Charlemagne as he sat on his throne, which was used for coronations throughout the Middle Ages. At midwinter noon, the Sun shines on a mosaic of Christ in Majesty. When Charlemagne stood up from his throne, he alone would have been able to see the equinoctial sunrise beaming horizontally through an upper window. For good measure, a sunbeam also shines on the throne on April 16th – Charlemagne's birthday.

At Elm, in the Swiss canton of Glarus, a natural rock tunnel, 60 feet (20m) in diameter, pierces the Tschingelhorner mountain that towers over the village to the southeast. According to legend, this tunnel, known as St Martin's Hole, was created when the saint hurled his iron-tipped staff at a giant. Each year close to the equinoxes, the morning Sun shines through this hole and for two minutes illuminates the tower of Elm church with a ray of sunlight just over 3 miles (5km) in length.

This event obviously predates the Christian era, but it is significant in terms of medieval sacred astronomy that the builders chose this spot for their church. In support of this interpretation is the fact that there are four more such holes in the Alps through which sunlight strikes a church.

SELECT BIBLIOGRAPHY

Atkinson, R.J.C. *Stonehenge*. London: Pelican (1979).

Aveni, A.F. *Skywatchers of Ancient Mexico*. Austin, Texas: University of Texas Press (1980).

Aveni, A.F. (ed.) *World Archeoastronomy*. Cambridge: Cambridge University Press (1989).

Brennan, Martin. *The Stars and the Stones*. London: Thames & Hudson (1983).

Burl, A. *Megalithic Brittany*, London: Thames & Hudson (1985).

Burl, A. *The Stone Circles of the British Isles*. New Haven: Yale University Press (1976).

Carlson, John B. "Romancing the Stone, or Moonshine on the Sun Dagger", in Carlson J.B. & Judge, W.J. (eds.) *Astronomy and Ceremony in the Prehistoric Southwest*. Maxwell Museum of Anthropology (1987).

de Santillana, Giorgio and Hertha von Dechend. *Hamlet's Mill: Myth and the Fame of Time*, Boston: Gambit (1969).

Devereux, Paul. *Secrets of Ancient and Sacred Places: The World's Mysterious Heritage*, London: Blandford Press (1993).

Devereux, Paul. *Symbolic Landscapes*, London: Gothic Image (1992)

Gauquelin, Michael. *Neo-Astronomy – A Copernican Revolution*, London: Penguin/Arkana (1991).

Graves, Robert. *The Greek Myths* (vols. 1 & 2), London and New York: Penguin (rev ed., 1960).

Hadingham, E. *Lines to the Mountain Gods: Nazca and the Mysteries of Peru*, London: Harrap and New York: Random House (1987).

Hawkins, G. with White, John B. *Stonehenge Decoded* (1973).

Hawkins, G. *Beyond Stonehenge*. London: Hutchinson (1973).

Hinckley, Richard and Dover, Allen. *Star Names: Their Lore and Meaning* (1963).

Hughes, David. *Star of Bethlehem Mystery*, London: Dent (1979).

Humprey, Caroline and Piers Vitebsky. *Sacred Architecture*, London: Duncan Baird Publishers (2003).

Kenton, Warren. *Astrology: the Celestial Mirror*. London: Thames & Hudson (1989).

Krupp, E.C. *Echoes of the Ancient Skies*. New York: Harper and Row (1983).

Krupp, E.C. *In Search of Ancient Astronomies*, Doubleday (1978).

MacKie, Euan. *Science and Society in Prehistoric Britain*. Elek (1977).

Malville, John McKim. "Astronomy at Vijayanagara" in *The Spirit and Power of Place*. National Geographic Society of India (1993).

Malville, John McKim. "Mapping the Sacred Geography of Vijayanagara" in *Mapping Invisible Worlds*. Edinburgh: Edinburgh University Press (1993).

Manilius. *Astronomica* (trans. G. P. Goold). Cambridge, Mass.: Harvard University Press (1977).

McMann, Jean. *Loughcrew: The Cairns*. After Hours Books (1993).

Michell, John. *A Little History of Astro-Archaeology*. London: Thames & Hudson (1989).

Millon, René. "The Place Where Time Began" in Berrin, K. & Pasztory, E. (eds.) *Teotihuacán*. London: Thames & Hudson (1993).

Molyneux, Brian Leigh and Vitebsky, Piers. *Sacred Earth, Sacred Stones*. London: Duncan Baird Publishers (2001).

North, John. *The Fontana History of Astronomy and Cosmology*, London: Fontana Press (1994).

O'Brien, Tim. *Light Years Ago*. Black Cat Press (1992).

Oppelt, Norman T. *Guide to Prehistoric Ruins of the Southwest*. Pruett (1989).

Pennick, Nigel and Devereux, Paul. *Lines on the Landscape*. Robert Hale: London (1989).

Ponting, Gerald and Ponting, Margaret. *New Light on the Stones of Callanish*. private (1984).

Ptolemy. *Tetrabiblios* (trans. F.E. Robbins). Cambridge, Mass.: Harvard University Press (1980).

Schafer, Edward H. *Pacing the Void – T'ang Approaches to the Stars*, California University Press (1977).

Sofaer, Anna., Zinser, Volker and Sinclair, Rolf M. "A Unique Solar Marking Construct" in *Science*, 206:4416 (19 October 1979).

Spence, Lewis. *The Gods of Mexico*. London: T. Fisher Unwin Ltd. (1923).

Staal, J. *Patterns in the Sky*. London: Hodder and Stoughton (1961).

Temple, Robert K.G. *The Sirius Mystery*. London: Destiny (1995).

Thom, A. *Megalithic Sites in Britain*. Oxford: Oxford University Press (1967).

Urton, G. *At the Crossroads of the Earth and the Sky: An Andean Cosmology*. Austin, Texas: University of Texas Press (1981).

Wilhelm, Richard (trans.) *I Ching or The Book of Changes*. London: Penguin/Arkana (1983).

Williamson, Ray A. *Living the Sky*, Norman, OK: University of Oklahoma Press (1984).

Wood, John Edwin. *Sun, Moon and Standing Stones*. Oxford: Oxford University Press (1980).

Select Bibliography

INDEX

Index

Index

Index

Index

Index

ACKNOWLEDGMENTS

The Publishers wish to thank the following individuals, museums and photographic libraries for permission to reproduce their material. Every care has been taken to trace copyright holders; however, if we have omitted anyone we apologize and will, if informed, make corrections in any future edition.

KEY: t: top, b: bottom, c: centre; l: left; r: right

AA: The Art Archive, London
AA&A: Ancient Art & Architecture
            Collection, Pinner
BAL: Bridgeman Art Library, London
BL: British Library, London
BM: British Museum, London
CWC: Charles Walker Collection, Yorkshire
MEPL: Mary Evans Picture Library, London
MH: Michael Holford, Loughton
V&A: Victoria and Albert Museum, London

Page 2 CWC; **11** CWC; **12–13** BAL/BL; **14** BAL/Bonhams; **44** The Pierpont Morgan Library/Art Resource NY; **46** CWC; **48** MH; **51** from *Gods of the Egyptians* (vol.1) by E. Wallis Budge, courtesy of the Trustees of the British Museum; **52** CWC; **55** CWC; **58** BAL/St John's College Library, Oxford, from *The Tables*, an accompanying volume to Dante's *Divine Comedy*, 1872; **61** MH; **66** AA/Bodleian Library, Oxford; **68t** Scala, Florence; **68–69** CWC; **70–71** AA/Sta Cruz Museo, Toledo; **72** AA/Trinity College, Cambridge; **74** Duncan Baird Publishers, London; **75** BM; **78** CWC; **79** CWC; **80** courtesy of the Maître de Chapelle de la Cathedrale de Saint-Lazare, Autun; **82** Israel Antiquities Authority, Jerusalem; **84** BAL/Bibliothèque Nationale, Paris; **87** AA/V&A; **89** BAL/BL; **91** CWC; **92** CWC; **96** CWC; **98** Ms. Stein 3326, Department of Oriental Manuscripts, BL; **100** CWC; **103** AA/Bibliothèque Nationale, Paris; **104–105** Collection Bibliothèque Municipale de Dijon; **106** AA/Biblioteca Marciana, Venice; **108** AA/Historika Museet, Stockholm; **110** BM; **112** AA, Egyptian Museum, Cairo; **113** CWC; **115** CWC; **116** CWC; **118** André Held; **119** Bildarchiv Preussischer Kulturbesitz, Berlin; **121** CWC; **123** BAL/National Museet Stockholm; **126** Werner Forman Archive, London/Liverpool Museum; **127** Scala, Florence/Museo del Terme, Rome; **130** CWC; **133** AA; **135** Master and Fellows of Trinity College, Cambridge; **136** BM; **140** CWC; **143** Royal Library, Copenhagen; **145** BM; **147** Scala, Florence/Museo Nazionale Reggio di Calabria; **150–151** CWC; **152** CWC **156** CWC; **157** CWC; **161** CWC; **163** BAL/Bibliothèque Nationale, Paris; **170–171** CWC; **173** CWC; **180** Ann Ronan Picture Library, Deddington; **181** JB Collection; **185** Ms Marsh 144, Bodleian Library, Oxford; **187** Ms Marsh 144, Bodleian Library, Oxford; **194** CWC; **195** AA; **199** BAL; **202** AA/Glasgow University Library; **209** BAL/Bibliothèque Nationale, Paris; **212** CWC; **221** AA/BL; **227** Ms Harley 647 f.4v, BM; **228** Museum of London; **233** CWC; **234** MEPL; **236** MEPL; **236–237** CWC; **238–239** Horizon, Guernsey; **242–243** Imagestate, London/Glover; **245** Imagestate, London; **256** CWC; **258** CWC; **267** Imagestate, London/Glover; **269** Mick Sharp; **278** CWC; **279** CWC; **289** Robert Harding Picture Library, London/Watts; **298** CWC; **300** MEPL; **302** Archaeological Institute of Kashihara, Japan; **305** John Collings **309** Paul Devereux; **313** Paul Devereux; **316–317** Imagestate, London; **324** Paul Devereux; **332** Robert Harding Picture Library, London; **333** Imagestate, London; **344** Robert Harding Picture Library, London/ Rennie; **337** Horizon, Guernsey/Walter; **338–339** Zefa, London; **348** AKG, London/Erich Lessing.

The Authors and Publishers would also like to acknowledge the following experts and researchers whose research or research, particularly on archeoastronomical sites, have been drawn upon for this book:

Eugene Antoniadi, Anthony F. Aveni, Prof. Alexander Badawy, Robert Bauval, Rev. Hugh Benson, Hiram Bingham, Charles Boyle, Martin Brennan, Stephan Peter Bumbacher, John B. Carlson, G. Charrière, Frank Hamilton Cushing, Michael Dames, David Dearborn, James Frazer, John M. Fritz, Leo Frobenius, Michel Gauquelin, Fred Gettings, John Glover, Gerald Hawkins, Willian Herschel, Fred Hoyle, David Hughes, Lucien Lévy-Bruhl, C.A. Newham, Sir Norman Lockyer, John McKim Malville, Terence Meaden, John Michell, Tim O'Brien, Michael J. O'Kelly, Gerald and Margaret Ponting, Richard Procter, Tom Ray, Ronald Lane Reese, F.S. Richards, Oliver Ricketson, Giorgio de Santillana, Anna Sofaer, Magnus Spence, Julius Stahl, William Stukeley, Wilhelm Teudt, Prof. Alexander and Archibald Thom, Dr. Virginia Trimble, Gary Urton, Herthes von Dechend, P. Wakefield, Hermann Weisweiler, John B. White, David Wilcox, Ray Williamson, Warren Wittry.

The Authors and Publishers would like to thank Sebastian Verney for permission to use an adaptation of his drawing on page 43.